REVOLUTION IN 24 HOURS

An AI-powered Journey
To Personal and Professional Excellence

SEBASTIAN VASQUEZ VILLAVICENCIO

Contents

CHAPTER 1
Waiting for Answers

Javier sat in the stark, sterile waiting room, dread pressing into his chest like a heavy fog. The television murmured in the background, its meaningless chatter failing to quiet his spiraling thoughts. He clutched the chair's metal edge, its chill biting into his skin—his only tether to reality. Each tick of the wall clock punctuated his anxiety, drawing him deeper into the unknown.

His thoughts spiraled in a relentless loop—his declining health, the pizzeria's struggles, and the ghosts of past failures whispering that he would never be enough.

The struggles at the pizzeria had recently intensified. A significant financial setback loomed over him—a missed rent payment gnawed at the already stretched budget. A steady decline in customer numbers, especially during peak hours, painted a dire picture. Ever since a new gourmet pizza chain opened nearby, even loyal patrons had dwindled. The landlord's reminders and the silence of empty tables compounded his anxiety, grounding his worries in harsh reality.

Beside him, other patients waited, absorbed in their own worlds of uncertainty, but Javier felt incredibly isolated. He vividly recalled every childhood doctor visit, each diagnosis and prognosis now echoing back with chilling urgency. Problems that once seemed distant now demanded immediate attention.

"Javier Lopez?" The nurse's voice sliced through his reflections, jolting him back to reality. With a deep breath, he rose, the weight of dread pressing down on his shoulders as he moved toward the doctor's office. The halls seemed unnaturally long, his footsteps hollow echoes of the questions swirling in his head.

In the office, Dr. Ruiz sat behind the desk, papers shuffled before him. Javier sat, his eyes flitting nervously around. The stark white walls adorned with framed certificates, a skeleton model in a corner, and the hum of fluorescent lights highlighted how removed he felt from his usual life.

Dr. Ruiz began speaking, his voice measured yet distant. "Javier, we have your test results... there's something we need to discuss." The words barely registered. A muted hum filled Javier's ears, drowning out everything but the pounding of his heart. The doctor's professional tone drifted past him, weightless, like a voice calling through fog. His gaze wandered to the books behind the doctor and the sturdy furniture, his heartbeat drowning out conversation details.

"See, the issue has progressed... your kidneys..." the doctor said, his voice cutting into Javier's consciousness. The specifics rushed back with clarity: "You'll need dialysis. Three times a week, four hours each."

The news hit like a physical blow, leaving him reeling. The room seemed to shrink, colors blending into white noise. He managed a nod, a gesture of compliance rather than understanding, barely registering treatment plans or assurances.

Stumbling back into the corridor, the clinical scent mixed with stale fear. His legs moved mechanically toward the exit. Outside, the bright sunlight felt harsh against his eyes, an altered reality crashing down.

Javier walked slowly toward the pizzeria, each step marking a new life chapter, shadowed by uncertainty yet tethered to survival's relentless drive.

As he walked, he started remembering his beginnings. Opening day at the pizzeria was a moment of triumph, forever etched in his memory. Morning sunlight gleamed against the freshly polished windows, illuminating the cozy interior he and his mother had prepared with painstaking care. Together, they had hung the "Grand Opening" banner across the entrance, her smile a beacon of pride. Customers had streamed in, drawn by the aroma of freshly baked pizzas wafting through the neighborhood. It was a day filled with hope, success, and the sweet promise of new beginnings, a testament to the resilience that Javier now sought to reclaim.

Suddenly, the pizzeria loomed ahead like a promise and a reminder, the family legacy buoying and burdening him. It was near midnight when Javier's phone buzzed, announcing an incoming Zoom call from Stephen, his college friend in New York. Javier joined, trying to mask his exhaustion.

"Hey, man, how are you holding up?" Stephen's familiar voice boomed, a comforting reminder of stress-free times.

Javier rubbed his temples, exhaling deeply. "I've been better, honestly. Had a rough day. Got some bad news at the hospital earlier."

Stephen nodded knowingly. "I heard. I'm sorry, Javi. But I've been thinking... maybe you should talk to someone about all this. You know, like a coach?"

"A coach?" Javier leaned back, skepticism knitting his brows.

"Yeah, my cousin was stuck too. A coach helped him sort through the noise and see things clearly. An ontological coach doesn't just give advice—they help you shift your mindset so you can actually move forward."

Javier considered Stephen's words. Structured guidance sounded appealing amidst the chaos. "What kind of coach are we talking about?"

"An ontological coach, perhaps. They manage your thoughts and align your inner self with your goals. I know someone who could fit. I'll send her info if you're interested."

A moment of silence hung before Javier murmured, "Yeah, maybe that would help. Send me her details."

They chatted longer, the conversation offering temporary normalcy. Signing off, Stephen added, "Remember, Javi, it's okay to slow down and focus on yourself sometimes."

After the call, Javier remained at his desk, staring blankly at the screen. Hope amidst uncertainty flickered, but problems couldn't be ignored. He opened a browser, searching for business ideas, focusing on AI integration into the pizzeria.

Amid endless articles, an all-in-one platform boasting AI solutions intrigued him. Streamlined operations and enhanced customer interaction promised to revolutionize his struggling pizzeria. He bookmarked it, envisioning potential benefits like automating online orders to handle peak times smoothly and crafting personalized marketing strategies to reach out to lost customers and bring them back, creating a more efficient and engaging operation.

The night wore on, Javier immersing himself in article after article about AI successes. Despite growing excitement, fear lingered—concerns about health, pizzeria survival, and his capability to meet challenges.

Exhaustion claimed him. Crawling into bed, thoughts of the path ahead swirled. With apprehension and determination, he resolved to explore new opportunities, hoping to convert fear into motivation. As sleep pulled at the edges of his consciousness, the allure of innovation battled deep-seated fears. The weight of past shortcomings and fresh health news fueled doubt. Staring at the ceiling, a new business dawn seemed dim under looming missteps. Battling these emotions, Javier decided to seek help, realizing he had to address his internal battles first. Determined to clear a path, he resolved to call Coach Laura first thing, hoping her guidance could navigate the murky waters of fear and uncertainty. Taking his diary, as he had since turning 21, he wrote:

> Tonight, the weight of everything feels heavier than ever. But I know now that asking for help isn't weakness, but survival. Strength comes from within, but clarity? That's the light others help you find when you can't see the way forward.

Little did Javier know that the call to Coach Laura would spark a journey that would challenge everything he thought he knew about himself and his business.

CHAPTER 2
The Weight of Legacy

Javier stepped into the dialysis center, his chest tightening with apprehension. The sterile air wrapped around him, thick with antiseptic and the mechanical hum of life-sustaining machines. Everything here was precise and calculated, a stark contrast to the unpredictable chaos of his world outside.

At the reception, the chief nurse greeted him with a professional but warm demeanor. She was an experienced woman with a no-nonsense attitude, projecting both authority and empathy.

"Good morning, Mr. Lopez. Welcome to the dialysis center. We'll be seeing a lot of each other," she started with a smile that was both reassuring and tinged with gravity. "Remember, your schedule is important. Arrive on time for each session—four hours, three times a week." She paused, ensuring he followed along.

Javier nodded, a heavy acknowledgment of his new commitment. "I understand. It's just... it all feels a bit overwhelming."

The nurse nodded sympathetically. "I know it's overwhelming, but we'll guide you through it. Since your kidneys can't function on their own, this machine"—she gestured toward the large cylindrical device—"will do the work for you. It will draw your blood, filter out toxins, and return it, repeating the process in cycles."

Javier listened intently, the weight of her words pressing into him. Anxiety curled tight in his chest, but her calm explanation reminded him of what he missed most—simplicity. The aroma of baked bread on a Sunday morning. The quiet companionship of a good friend. Life before everything became complicated. Each thought brought a fleeting smile, a reminder of life's little joys.

"Okay, let's get you started. Sit here, please." She gestured toward the chair beside one of the machines, the centerpiece of his sessions for the foreseeable future.

He took the seat, cool leather sending a shiver up his spine. As he settled in, the nurse prepared everything with practiced efficiency. "You'll feel a little pinch now," she forewarned, gently taking his arm.

Javier turned his head, bracing himself. The needle pricks were sharper than anticipated, jolting him more than the nurse had implied. A quiet gasp escaped him, betraying the surprise of pain against his rigid composure.

"Alright, all done," she said, her voice steadying. "Now, you've got four hours to relax. You can watch TV, study, anything you like."

Javier gestured to his bag. "Mind if I use my computer?" "Of course," she nodded, her approval granting him a sense of normalcy amidst the chaos of adjustment.

With trepidation, Javier opened his laptop. He connected to the pizzeria's bank account, a routine task that now seemed daunting. As the figures loaded, his stomach twisted tighter than the pinch's residual pain. The numbers, stark and unforgiving, painted a dire picture, each digit a testament to his mounting fears. Missed rent payments loomed on his spreadsheet. The numbers were unforgiving—customer visits had steadily declined while ingredient costs climbed. The weight of reality settled in, sharp and relentless.

The further strain comforted little. That moment of routine, the edge of the financial absorbed working capital, was difficult to change. In financial terms, the burden of new life's uncertainties and financial both machine to a tether felt heavier than ever.

The hum of the machines filled the air like an unending drone, punctuated occasionally by quiet murmurs between patients and the medical staff. Javier observed the room, taking in the twelve stations, each occupied by someone tethered to the life-preserving apparatus. It was a community bound by necessity, yet everyone was ensconced in their own world of thoughts and hopes.

Amidst this solemnity, Dr. Wood made his rounds, a figure of reassurance moving from patient to patient. Eventually, he stopped beside Javier, who looked up, meeting the doctor's calm gaze.

"Hi, Javier. I'm Dr. Wood. I'll oversee your treatment here." His voice carried a gentle authority that momentarily eased the knot in Javier's chest. "There are some key guidelines you'll need to follow: limit your water intake to one liter a day, cut out added salt, and stick to a low-potassium, low-phosphorus diet. And be careful with flour products—they can be risky."

Dr. Wood handed Javier a printed list, breaking down his dietary restrictions in detail. As Javier scanned the document, a sense of disbelief shadowed his features. Everything on the list seemed to include all his favorite foods. Silent, he nodded in acknowledgment, but internally, his thoughts spiraled.

"What about the food at the pizzeria?" His mind raced. The tantalizing aroma of freshly baked dough, the vibrant flavors of his special sauces, and the gooey, comforting bite of a well-made pizza—all suddenly felt off-limits. How could he manage his role in the business without tasting the food he was passionate about? Could he truly guide the pizzeria to success without engaging his senses completely?

As this thought occupied him, Javier's memory flickered back to a simpler time—a bright morning in the pizzeria's bustling kitchen. Flour dust danced in the warm air as he and his team experimented with a new recipe. He remembered the triumph of perfecting a sauce that was met with rave reviews. The joy of watching satisfied customers savor each bite had filled him with pride. Back then, he had felt unstoppable, his creativity boundless, his passion unshakable.

Yet here he sat, in stark contrast, grappling with the boundaries now imposed upon him. The taste of yesterday's victories seemed farther away than ever.

The questions piled upon his existing worries, like waves in a relentless storm. As Dr. Wood moved on to the next patient, Javier felt the weight of these new restrictions settle heavily upon him, tightening the web of anxiety spun from his dual challenges.

Looking at his laptop, the screen still displayed the harsh realities of his financial predicament. The juxtaposition of medical constraints and business hurdles pressed down on him, an almost physical force. The weight of his fears, his failing health, and the pizzeria's struggles crashed down on him—suffocating, unbearable.

With a swift motion, Javier closed the laptop, shutting out the crisis looming for a moment of reprieve. He fished out his iPhone; the familiarity of its weight grounding him slightly. Music, his refuge and escape, filled his ears as the first notes played. He leaned back, allowing the melody to wash over him, temporarily blotting out the cacophony of concerns.

Amidst soothing sounds, his heart wretched with a deep-seated vulnerability over fear, his health, the pizzeria's future, and his ability to reconcile the two loomed large. Despite his outward silence, an inner dialogue surged forth—a determination to not let the burdens defeat him. As the session wore on, Javier steeled himself, knowing that while he felt beaten, surrender was not an option. Each song became both a balm and a challenge, urging him to find his way through this murky path. Despite the chaos, a smile tugged at his lips as a memory nudged its way to the forefront—a memory that never failed to lighten his heart.

It was a particularly frenzied Friday night, and the pizzeria was a tempest of voices and orders. Javier, caught in the whirl of tossing dough and pepperoni, had thrown a pizza with such enthusiasm that it soared upwards, only to land perfectly centered on the ceiling. The entire kitchen froze, eyes wide with surprise, then burst into a cacophony of laughter. Even the stern chef couldn't resist the humor, announcing with mock gravity that Javier's creation was "a heavenly offering to the pizza gods." The recollection of that night, with its shared smiles and camaraderie, reminded Javier that even in the midst of life's pressures, there was always room for laughter. It was moments like these that fortified him, underscoring that resilience was sometimes built one chuckle at a time.

Returning home from the dialysis center felt like releasing a tightly held breath, yet it offered no relief to Javier.

Upon opening the door to his apartment, Javier was hit by the lingering scent of burnt cheese from the night before, a stubborn reminder of his rushed attempt at dinner. It mingled with the cool, sterile air that clung to his clothes from the dialysis center, creating an uncomfortable blend of domesticity and clinical detachment. The once cozy living room now felt oppressively small, its cluttered corners hemming him in like a trap. The hum of the refrigerator was the only sound, an intrusive buzz against the stifling silence.

The soft glow of the streetlight seeping through the worn curtains cast long shadows, mirroring the tangled thoughts in Javier's mind. He rubbed his temples, feeling the grit of exhaustion in his bones, yet the soft texture of the couch beckoned him to find a moment of peace. He sank into it, closing his eyes, trying to drown out the oppressive sensation of familiarity that both comforted and suffocated him.

The walls of his apartment seemed closer than usual, clawing at his suffocating sense of inadequacy. Javier was immediately consumed by the weight of unanswered questions, his mind a tumultuous sea of what-ifs and lingering insecurities.

He moved through his evening rituals with a mechanistic detachment, his actions void of their usual purpose. As he slipped into bed, the shadows of his room only emphasized the darkness he felt within. Javier's thoughts were a relentless spiral, pulling him further into an abyss of uncertainty. He had always prided himself on his resilience, yet tonight that very reserve felt depleted.

The enormity of his ordeal—dividing time between his health demands and the critical needs of the pizzeria—pressed down heavier still. "Can I really do this?" he thought, his doubt echoing in the silent room. The thought of cutting back his time at the pizzeria to accommodate the dialysis sessions was daunting. He felt the creeping dread of losing control over everything he held dear.

Desperate for connection, Javier reached for his phone. The need to hear a familiar voice, a lifeline, was overwhelming. "Stephen, sorry to bother you, man, but I need help. I started dialysis today, and..." His voice wavering, he confessed, "I'll have like 12 hours less to be at the pizzeria. I'll have to work online to make it..."

Stephen's voice, warm and steady, emanated from the speaker. "Javi, I get it. That sounds tough," he said, pausing to let his words sink in. "But remember, you're not alone in this. I know things seem impossible right now, but having someone to guide you can make all the difference."

Javier hesitated, his pride wrestling with his vulnerability. "Do you really think that coach could help?"

"I really do. Coaches are trained to help untangle the knots in our lives, offering not just solutions but a new way to look at the problems." Stephen leaned back in his chair, his eyes reflecting sincerity and warmth. "You know," he began, "my cousin was once in a tight spot, financially speaking. Drowning in debt and unable to see a way out, he felt like everything was closing in on him. It was a coach who stepped in, though. They didn't just offer generic advice; they helped him craft a detailed plan. They set small, achievable goals that slowly but surely pulled him out of the red. Today, he's not just surviving—he's thriving, with a life he never thought possible." Stephen paused, letting the weight of the story settle. "Sometimes, a little guidance is all it takes to turn things around."

Javier listened, a part of him yearning to believe. Stephen continued with comforting conviction, "It's worth a try, right? Even if it just helps you handle things a bit easier."

Feeling the weight of Stephen's supportive insistence, Javier managed a slight nod, as if his friend could see. "Alright, I'll give it a shot," he murmured, both to Stephen and himself.

The call ended, leaving Javier once again with the quiet shadows of his room. But now, accompanied by the lingering warmth of their conversation, there was a flicker of something else—a resolution, fragile yet present.

As he lay back down, the familiar dread still hovered, yet it had lost some of its menace. Javier's thoughts danced between doubt and determination; the balance ever so slightly tipped towards the latter. With a whispered promise to himself, he resolved to meet the coach, diving into research the next day to find the best fit and reorganize his schedule to carve out time for both his health regimen and managing the business.

He opened his notepad and began jotting down ideas for untraditional topping combinations and cost-effective sourcing strategies—something to breathe new life into the pizzeria. Each stroke of the pen felt like a step forward against the tide of challenges he faced. He envisioned new menu items that would not only attract customers but also set the pizzeria apart in an ever-growing sea of eateries.

This day had been full of things he needed to order, so he took his diary and wrote: "Tonight feels like the end of a relentless chapter—one where confusion and exhaustion held me hostage. But today has made me realize how easily life's pressures can cloud my vision. Through the darkness, I've begun gathering fragments of clarity. Tomorrow, I'll start researching coaches, drafting a plan, and carving out a routine that lets me fight for both my health and the pizzeria. One step at a time, I'll reclaim control. I'll find my way."

With that written, Javier felt a slight but significant shift in his perception. Tomorrow was a new day, full of challenges but also hope, a step towards reclaiming control and finding his way through life's complexities.

CHAPTER 3
Stepping into Uncertainty

Javier sat in the hushed waiting room outside Laura's consultation office. The rhythmic ticking of an old-fashioned clock punctuated the silence, each tick a stark contrast to the chaos swirling in his mind. Every second stretched, amplifying the weight of why he was there.

Despite the room's serenity, unease coiled in his chest. His thoughts darted from the crushing weight of his diagnosis to the ghosts of failed ventures, lingering like unwelcome companions. Every misstep, every broken plan wove itself into a suffocating tapestry of doubt.

As he sat there, images of how the pizzeria had barely managed to scrape by over the last year dominated his thoughts. What had once been a beacon of hope—a family legacy he dreamed of transforming—now felt like an enormous weight. The prospect of dialysis loomed large over everything, a reminder of his own fragility and the pressing need to secure his future.

"Javier?" A warm voice broke through his spiraling thoughts. He looked up to find Laura standing at the door, her inviting smile a quiet reassurance. Something about her presence steadied him—like an anchor in the storm.

"Hello, I'm Laura," she greeted, gesturing for him to follow her inside. Javier nodded, managing a polite smile as he rose and followed her in.

Inside, the office was a cozy, well-organized space. The walls were lined with bookshelves and dotted with framed certificates and diplomas that attested to Laura's expertise and commitment. A large window let in streams of natural light, alongside potted plants that added a touch of life to the room. This should have been reassuring, yet Javier couldn't help but feel a renewed wave of pressure.

As he lowered himself into a comfortable chair opposite Laura, he allowed his eyes to wander around the room. He couldn't help analyzing every detail: the arrangement of books that suggested both order and warmth, the array of diplomas that silently vouched for Laura's capabilities. But instead of feeling at ease, his thoughts only spiraled further.

Despite the elegant order around him, he found himself overwhelmed by a sense of inadequacy. His gaze fixed on an abstract painting—a chaotic swirl of color, mirroring the storm inside him. Laura's calm presence barely touched the battlefield of uncertainty raging in his mind. "Why am I so stuck?" he wondered. Each new fear sparked another, creating a cascade of hypothetical disasters. He worried about the future of the pizzeria and how he could manage it while handling his health. Every potential solution felt drowned by the shadows of doubt.

Laura took her seat, her demeanor gentle and patient. Sensing his discomfort, she leaned forward slightly, allowing her genuine interest to seep through the space between them.

Lost in his swirling doubts, Javier almost missed Laura's gentle voice breaking through, grounding him back in the moment. "Javier, why don't you start by telling me a bit about yourself?" she invited encouragingly.

Caught in his own storm, Javier paused. He took a deep breath, preparing to delve into the tangled web of his life, hoping that by putting it into words, he might just begin to untangle it.

Laura leaned forward slightly, meeting Javier's weary eyes with genuine empathy. "Javier, I want to first acknowledge how difficult this must be for you. It's clear you're going through a lot, but seeking help like you've chosen to do today is a brave and crucial step toward change. I'm glad you're here."

Javier nodded, a slight tremble in his voice as he replied, "Thank you, Laura. It feels... overwhelming, but I knew I needed to talk to someone."

Laura continued, a gentle seriousness in her tone. "Let's start by ensuring that your medical situation is managed properly. Javier, are you following up with all your medical appointments regularly?"

"Yes, I am," Javier confirmed. "I make sure to attend all necessary appointments, though sometimes it feels like it's all I do."

"That's important. And are you following the prescribed treatment plan for your dialysis? I understand it can be incredibly taxing," Laura asked, her eyes reflecting concern.

"I am," said Javier, nodding slightly. "Three times a week, four hours each, like the doctor recommended. It's tough, but I'm sticking with it."

Laura observed him thoughtfully for a moment before asking, "How do you feel about this routine? Is there support in place for you outside of these sessions?"

Javier took a deep breath, considering his answer. "It's draining, both physically and emotionally. My mom and a few friends check in, but it often feels like I'm alone dealing with it all."

Acknowledging his isolation, Laura posed another question, "And in terms of managing stress and balancing everything else, do you find any time to focus on activities that help you recharge?"

Javier hesitated. "Not really. The stress... it takes over. Some days, I forget what it feels like to breathe freely."

Laura nodded slowly, understanding the gravity of his statement. "It sounds like you've got a lot on your plate, Javier. But you've already taken a huge first step by coming here. Recognizing these gaps is the first step to finding solutions that can help. We can work through them together."

In the quiet that followed, Javier absorbed Laura's empathic support. The dialogue had drawn a faint line of hope through the complex tapestry of his current struggles.

Javier glanced at the clock on the wall, surprised to see that an entire hour had slipped by. The session had flown by, the weight of his worries lightened slightly through sharing. Just then, Laura's voice broke into his thoughts.

"It seems our hour is up," Laura said gently, her tone both concluding and inviting. "Before we finish, I'd like to ask, what are you taking home with you today, Javier?"

Javier leaned back in his chair, considering her question carefully. "I think... I'm taking home a bit of perspective, honestly. Talking things through has made me realize how much I've been keeping inside and how it's been affecting me more than I acknowledged."

He paused, gathering his thoughts. "And maybe a sense that I'm not completely alone in this," he added. "Knowing that there's a way through and that I can look to someone to guide me, even if it's just to find clarity in little pieces. It's a start."

Laura smiled warmly, nodding in approval. "That's good to hear, Javier. Remember, each step counts, no matter how small. I'm here to support you as we work through this."

As he stood to leave, Laura offered a final piece of advice. "Javier, to achieve any goal you have, we are going to learn to analyze a situation, understand our feelings, and modulate them so they start being manageable for you." She paused, ensuring he caught the weight of her words. "We will focus only on the problems where you have a space to change something."

Javier nodded, digesting her insights with a newfound sense of purpose. Laura called after him as he walked towards the door, "See you in two weeks."

With those words, Javier left his first session, stepping out with more than just advice; he carried a sliver of hope toward his unfolding journey.

A New Day, New Challenges

Javier settled into his cramped office space, his small haven amidst the chaos of the pizzeria. The room functioned as a storeroom as much as it did an office; boxes of supplier goods were neatly stacked against the walls, leaving just enough room for his desk and chair. Despite its size, Javier found a certain solace in this space, filled with the tangible reminders of operations and growth.

The morning light filtered through a single window, casting a warm, golden hue over the scattered papers on his desk. Beyond the glass, Javier could glimpse the restaurant floor. Only three tables were occupied, and each group of patrons represented a small victory in an otherwise challenging landscape. The sight of empty tables, however, was a stark reminder of the quiet desperation simmering beneath the surface.

His desk was a minefield of financial documents—bills, invoices, inventory lists—a paper trail of the pizzeria's current struggles. The recent spike in supplier costs loomed large on his computer screen, figures that seemed to grow in their taunt with every glance. The rising cost of mozzarella alone had eaten into their already slim margins, forcing Javier to contemplate cutting back on portions.

Javier took a deep breath, the air heavy with the scent of flour and tomatoes wafting from the kitchen. He felt the familiar churn of anxiety knotting in his stomach, a sensation that had become all too frequent. "How can I turn this around?" he wondered, the question echoing the thoughts that kept him awake at night.

He leaned back, rubbing his temples, trying to stave off the frustration building within. His mind flashed through memories of sleepless nights spent calculating expenses, the constant stress of uncertainty wrapped tightly around his chest.

In these moments, the enormity of his predicament often felt suffocating. His thoughts flitted to his father, imagining how he might have weathered such challenges. Was Javier living up to his legacy? Had staying true to the family business been the right choice?

As these thoughts clashed in his mind, he recalled Laura's words from the day before, words that had rung with a truth he had yet to fully grasp. "Focus on what you can control," she had advised, cutting through the clutter of his anxieties with a clarity that now offered him a lifeline amidst the storm.

Rallying himself, Javier leaned toward the computer, intent on finding solutions rather than dwelling on problems he couldn't immediately solve. He began delving into potential innovations, considering ways to integrate modern technology into his traditional business.

His research led him down varied pathways: exploring AI's role in enhancing customer engagement, the promise of an AI-driven menu that could adjust in real time based on customer preferences sparked a glimmer of possibility, optimizing supply chains, and predicting trends. Each innovation promised a realm of possibilities that, while intimidating, also excited him with its potential. The tools showed a clear pathway to a more personalized experience.

With every article and case study, Javier's vision broadened. He imagined how such tools could breathe new life into the pizzeria. The thought of AI-driven efficiencies lifting the heavy burden he felt was increasingly tantalizing.

This exploration transformed his office from a repository of stress to a haven of hope. Javier began dreaming aloud of dynamic menu updates that could entice the lunch crowd or customized marketing campaigns that could draw in weekend diners. The possibilities seemed as limitless as the data algorithms promised.

Yet, the underlying fear of implementing such changes without guaranteed success still lingered, as persistent as ever. Javier knew he'd have to face obstacles head-on, shouldering risks in a field where hesitancy had no place.

Stepping back from the immediate dread of bills and costs, he started understanding Laura's counsel: to channel his energy where he could effect change, to not drown in what remained beyond him. It wasn't about discarding the challenges but rather navigating through them with strategic clarity.

With each new idea, the landscape of the pizzeria's future started shifting subtly. Though uncertainties remained, Javier felt bolstered by the prospect of progress. There was a budding determination to guide the business—and himself—toward a vision not yet fully formed but enticingly possible.

As he sat back, exhausted yet resolute, the afternoon light spilled in, casting long shadows across the boxes. Javier knew that transforming dreams into reality was a journey laden with trials, but for the first time in a long while, he felt that it was a journey worth undertaking.

Javier closed his eyes for a moment, allowing the restaurant's buzz, the distant chatter of patrons, and the digital glow of his research to merge into a single, coherent stream. A future reimagined—one where he steered with renewed intent and insight.

As the day's tasks beckoned, Javier felt the pull of purpose stronger than the weight of doubt. Inspired by thin yet steadfast threads of possibility, he faced the path ahead with newfound courage to reshape the narrative of both his life and business.

As the afternoon shadows began to stretch longer across Javier's office, he continued his deep dive into technological solutions. Each mouse click opened new possibilities, though the sheer volume of information often felt overwhelming. Just as he was about to call it a day, a notification drew his attention to Instagram.

Just as he began feeling overwhelmed by the complexity of the options, a notification popped up, offering a promise of clarity—an invitation to watch Mike Rubence's video.

Curious, he clicked on a video titled "Revolution in 24 Hours: Boost Your Business with AI." The bold promise was both intriguing and intimidating. As the video loaded, Javier leaned in, his interest piqued.

The screen flickered to life with the charismatic presence of Mike Rubence, who greeted viewers with infectious enthusiasm. "Hello, entrepreneurs and innovators! I'm Mike Rubence, the CEO and Founder of MKTGAIPRO," he announced, the confidence in his voice undeniable. "As a platform trusted by hundreds of small businesses, Mike's confidence was backed by a proven track record. We've developed the ultimate AI platform designed specifically for you—those who refuse to be left behind and are ready to dominate their market!" This platform promised to simplify operations, improve customer engagement, and reduce costs—all tools Javier desperately needed.

Javier felt a twinge of excitement at Mike's words. Here was someone offering exactly what he had begun to seek: transformation, efficiency, and growth through cutting-edge technology.

"Imagine revolutionizing your business in just 24 hours," Mike continued, his tone persuasive and empowering. "With MKTGAIPRO, you gain access to the coolest apps that not only enhance operational efficiency but also provide a competitive edge that can help you crush your goals and leave the competition lagging behind."

Javier's heart quickened with each promise; the vision of transforming his struggling pizzeria into a thriving enterprise suddenly seemed within reach. The energy and certainty in Mike's voice were contagious, igniting a spark of optimism within him.

Mike leaned slightly forward in the video, eyes glinting with passion. "Don't let fear hold you back! The future is here, and it's yours to shape. With strategic AI integration, you can unlock potential you never even realized was possible. It's time to step up and own your business success."

The pitch resonated deeply with Javier, stirring a whirlwind of thoughts about possibilities and opportunities. The idea of leveraging AI this way felt like seeing a map to a treasure he had sought for years.

Yet, as the video ended, a shadow of doubt crept back in, unwelcome but persistent. With all this potential came the weight of expectation and the familiar fear of failure. Could he really navigate this new terrain without stumbling?

Javier sat back in his chair, the flickering light from the screen casting shifting patterns across the scattered documents on his desk. The excitement from the video was undeniable, but so was the reemergence of his old insecurities. He knew he needed to tread carefully, to balance aspiration with pragmatism, ensuring he didn't lose sight of the stability he craved amid the chaos of discovery.

As the office dimmed with the setting sun, Javier remained thoughtful, caught between the electrifying prospects of a bold tomorrow and the haunting echoes of yesterday's doubts.

As he finishes the day, he throws himself onto his bed, grabs his diary, and writes: "Tonight, as I close my eyes, I'm struck by the strength that lies in simply voicing your fears and ambitions. Walking into Laura's office had felt daunting, but it sparked something I didn't know I needed. Today taught me the art of discovery—how facing the past and aligning it with the present can forge a resilience stronger than doubt." The obstacles now seem less like insurmountable walls and more like puzzles waiting to be solved. I find solace in this new self-awareness, the power unlocked through deep introspection and strategic action. The discovery of a promising AI feature that could predict demand and invent activities for their preferences was a capability that could bring in revenue, creating a great illusion. Tomorrow, I'll start mapping out the first steps to integrate these new ideas into the pizzeria—one piece at a time.

CHAPTER 4
A Path Unfolds

Javier sat across from Laura once more. The familiarity of her consultation office was both a comfort and a stark reminder of the crossroads before him. Sunlight streamed through the window, painting the room in shifting patterns of gold—but the warmth did little to untangle the knot of anxiety tightening in his chest.

Caught between the pizzeria's struggles and his own health, every decision felt like a potential misstep.

Laura observed him patiently, her demeanor a consistent well of calm that Javier found grounding. Sensing his internal chaos, she leaned forward slightly, her voice carrying a gentle power. "Javier, I can see you're carrying a lot. But let's focus on what you can actually control—one step at a time."

He sighed, running a hand through his hair, feeling the pressure of indecision press down on him like a physical weight. "I just don't know what to do next, Laura. It's like... every option feels like it could either save or ruin everything."

Her eyes lingered on his face for a moment before she asked a simple question that caught him off guard: "What is in your hand to do in the present situation?"

Her words lingered, cutting through the fog of uncertainty. What was truly within his grasp? He sifted through the overwhelming possibilities, searching for something solid—something he could act on now. What was in his hand right now? What could he control? It was a gentle reminder to focus on the immediate, actionable elements of his life that he could influence directly without spiraling into despair over the vast, uncontrollable uncertainties.

Javier paused, letting the question sink in. He realized that amidst all the chaos, there were indeed threads he could grasp: the pizzeria's day-to-day operations, the possibility of gradually integrating new technologies, and the conversations he had yet to have with potential partners or mentors. Laura's question carved through the noise, guiding him back to the grounding simplicity of taking small, actionable steps.

He nodded slowly, the knot in his chest loosening ever so slightly. "I guess... I can focus on small changes—things I can do today instead of trying to fix everything at once."

Laura smiled, her eyes alight with encouragement. "Exactly. The biggest transformations often start with the smallest shifts. Focus on those, and the path ahead will come into view." As Javier sat in the quiet warmth of the office, he felt a shift within himself. The crossroads didn't seem as intimidating as before; the weight of the decision was not as paralyzing. With Laura's guidance, he sensed a way to navigate his challenges with intention and care, taking one step at a time.

Laura leaned forward, continuing her questioning. "Javier, what are your best qualities that can help you in your business?" she asked, the inquiry designed to help him recognize his strengths.

Javier paused, considering the breadth of his experiences, and began, "Well, technically, I've got a solid foundation. My education gave me an MBA and a diploma in marketing, both of which have equipped me with strategic insights. I've been immersed in the business side of things for so long, managing not just the pizzeria but also tackling challenges in my previous ventures. I'm analytical, always looking to optimize operations and improve systems, and I've got this vision of where I want the business to go, especially with the integration of technologies like AI."

As he spoke, Javier's voice grew more confident, buoyed by the recognition of his capabilities. He delved into specifics: his ability to draft business plans, execute marketing campaigns, and apply his analytic skills to drive growth. Each point underscored a capability rooted in his professional life.

Laura listened carefully, nodding occasionally, her silence encouraging him to elaborate further. Once he finished, she offered a thoughtful smile before proposing another angle: "Okay, those are impressive qualifications. But how do you combine all this with your personal background?"

This question subtly shifted Javier's demeanor, prompting deeper reflection. It was an invitation to examine how his life story, personal trials, and triumphs intermixed with his professional attributes, enriching them with emotional integrity and authenticity.

Javier hesitated, then replied, "Well, personally, I suppose my upbringing has taught me resilience and perseverance. Losing my father early and growing up with the responsibility of helping my mom with the pizzeria instilled in me a deep sense of commitment and family values. I think this translates into a leadership style that's empathetic and driven by integrity."

He continued, "Being diagnosed with type 1 diabetes forced me to be disciplined and attentive to details, akin to how I run the business. And, I guess, my life's emotional highs and lows have given me a certain depth in understanding customer needs and relations because I focus on connecting with people."

Laura nodded, sensing a fuller picture emerging. "Integrating these aspects—your technical expertise with your personal lessons and values—will strengthen your approach, Javier. Your business decisions will not just be informed by strategic knowledge but also by a human touch, which is invaluable."

Javier considered her words, recognizing the potential of blending his learned skills with his deeper emotional experiences. The integration of these layers promised a robust business strategy and a leadership approach that could inspire genuine connection and resilience within his team and with his customers.

Javier was animated as he explained his vision of integrating AI into the pizzeria's framework. "By utilizing AI, I can streamline operations, enhance customer service with personalized experiences, and predict trends in what customers might want next. Imagine tailored promotions that bring regulars back more often, or even just optimizing supply orders to cut down on waste and improve margins," he continued, a note of excitement threading through his voice.

Laura listened closely, a smile tugging at the corners of her lips. Once Javier concluded, she quipped with a playful tone, "So, you would be like a Terminator?"

Javier chuckled, puzzled yet amused. "Why's that?" he probed.

"Well, think about it," Laura replied, her eyes twinkling with humor and sincerity. "You'd be a sort of cyborg—your knowledge of business and strategy enhanced with cutting-edge technology. Plus, you have what I consider one of your best qualities: resilience. You resist more than a lot of people, and you keep going."

Javier couldn't help but grin at the comparison, finding it oddly apt. It struck a chord with him because it framed his strengths in a light he hadn't considered before. "I suppose persistence is one of my stronger suits," he acknowledged, gratitude in his voice.

Laura continued, "Exactly! And like the old song goes—the one you just hinted at—when the going gets tough, the tough get going. That's you, Javier."

The words resonated deeply within him, stirring a wellspring of inspiration and determination that had been lying dormant. Having someone else recognize his resilience validated feelings he had long kept at bay due to self-doubt and the fear of failure. He could feel a shift happening within, implanting a robust seed of newfound belief in his own capabilities.

The session ended, Laura's affirmations lingering in the room like an embrace that provided encouragement and reassurance. As Javier stepped out into the crisp afternoon air, he felt lighter, as if Laura's office had been a chrysalis and he was emerging renewed.

His walk home was not just a physical journey but an introspective one. Thoughts of how he could implement AI into the pizzeria played in his mind like a new melody. Ideas for innovative customer engagement and efficient management filled him with excitement rather than dread.

Javier reflected on the conversation with Laura. Her metaphor of the Terminator—a blend of human intuition and technological advancement—epitomized the potential pathway for himself and his business. He realized that concerns about past failures did not hold power over him unless he allowed them to. Instead, integrating his past lessons with promising new tools could forge a future unlike any he had dared to dream of.

As he continued walking, he absorbed the sounds of the city around him, each step firm with purpose. Quiet confidence replaced earlier hesitations, and he embraced the metaphorical duality Laura had pointed out. His resilience and adaptability, combined with intelligent innovation, could truly transform the pizzeria into a modern marvel of efficiency and customer satisfaction.

Javier's heart soared with the possibilities. A vision formed: a bustling pizzeria thriving not just on tradition but on the seamless weaving of tradition and technology. He could see it as a hub of satisfied customers, as familiar orders were made with unprecedented precision, and new dishes enhanced by data-driven insights delighted guests.

The promise of a revitalized business, supported by the strengths he already possessed and the innovations he now sought to embrace, filled him with hope. He imagined his father watching with pride, his legacy not only preserved but elevated to new heights through Javier's tenacity and vision.

By the time he reached his doorstep, Javier's mind was buzzing with plans and the certainty that he could indeed reshape his narrative. Laura's words about his toughness and persistence became a mantra he recited softly to himself, solidifying his resolve.

Entering his home, Javier knew that the road ahead, though challenging, was one he wouldn't traverse alone. Armed with clarity, determination, and a touch of humor at the thought of being a modern-day cyborg, he was ready to face whatever lay ahead with courage and conviction. This truly felt like the beginning of an exciting new chapter.

A Midnight Revelation

The soft blue light of the computer screen cut through the darkness of Javier's office as midnight ticked past. The stillness of the night wrapped around him, punctuated only by the faint hum of his computer. He was about to shut it down when an advertisement appeared, Mike Rubence announcing a special discount on the MKTGAIPRO platform, plus an exclusive course. The offer was timely, and Javier's interest was piqued.

His gaze wandered to the old photograph on the wall, capturing his father in uniform—a stern figure, the embodiment of discipline and duty. A flood of memories surged as Javier looked at the photo, recalling the distant relationship they had shared. For his father, a military man through and through, affection was shown through tough lessons rather than tender words.

In his mind's eye, Javier was a young boy again, sitting rigidly across the dining room table from his father. The ambiance was one of strict discipline; his father demanded nothing less than perfection, whether in school or in understanding the precise movements of military maps. It was during one of these lessons, amidst stacks of paperwork and tactical diagrams, that he had learned the importance of precision and order, though it rarely came with words of encouragement. He remembered the dining room table strewn with tactical diagrams, his father's sharp voice slicing through the quiet as he corrected Javier's every hesitation.

His father, though unsentimental, instilled within Javier a powerful work ethic and a resilience seen only in the face of challenges. "You need to be strong, Javi," his father would say, the words sharp as the creases in his uniform. "Life won't hand you victories. You earn them."

Javier absorbed these lessons with a mix of respect and longing for the warmth that never came. The photograph wasn't just a relic of the past—it was a mirror of the values shaping his own path. His father's discipline had trained him for moments like this: to assess, decide, and act purposefully.

With a deep breath, he returned his focus to the digital offer on his screen. Integrating AI into his pizzeria was more than just a business move—it was an evolution, a chance to bring order to the chaos that threatened his livelihood. It was a way to honor the disciplined approach his father had always demanded while carving out his own path forward.

As he clicked the link, Javier felt a surge of determination. This wasn't just a purchase; it was a commitment to a new way forward.

While Javier's relationship with his father was not marked by warmth, the strength of character that was nurtured in him through those early lessons was undeniable. They were the foundation upon which he could build, adapt, and succeed, even when it meant venturing into uncharted territories like AI technology.

With the deed done, he leaned back and allowed a moment of reflection. The transaction wasn't just about a software platform; it was about embracing a future filled with possibility while still rooted in the resilience forged under his father's watchful eyes. The night's quiet bore witness to a son who was ready to blend the clarity of a soldier's lessons with the creativity of an entrepreneur's spirit.

When he returned home tired, he sat on his favorite sofa in the living room and saw his diary and pen. He immediately took them and wrote: "Lying here in the stillness, reflecting on today, the convergence of past and future paths unfolding within me becomes wonderfully palpable. The midnight hour brought clarity and realization during my interaction with Mike's platform that my health and business struggles are intertwined stories awaiting their new chapters. It unveils the lesson: harnessing inherited resilience and pairing it with cutting-edge innovation can chart an incredible course. I've seized a moment to commit to blending old lessons with fresh opportunities, ensuring my forward journey is rooted in rich potential. Tomorrow, I'll map out the first steps for integrating AI into the pizzeria, starting with customer feedback tracking."

CHAPTER 5
The Allure of Innovation

As the first light of Saturday morning filtered through his window, Javier was already at his desk, his pulse quickening with anticipation. He logged into the AI platform he had just purchased, eager to explore its vast potential. More than 20 apps gleamed on the dashboard—each a tool capable of transforming his pizzeria, streamlining marketing, and drawing in new customers. The sheer scope of possibilities left him awestruck.

The first category he explored was explicitly labeled "Marketing Apps." Here, he found options to automate campaigns and strategies he had only previously imagined. There were tools designed to pinpoint leads accurately and nurture them through sophisticated, automated sequences, all crafted to maximize sales and visibility.

He clicked on the app labeled AI Videos. A demonstration video popped up, showcasing dynamic animations and personalized video messages that could be created on the fly. Javier marveled at the simplicity—input a few key details about a promotion or menu special, and the AI would conjure an engaging video complete with voiceovers and captivating visuals within minutes. His thoughts raced with possibilities, envisioning how these videos could captivate customers on social media platforms or within email campaigns. He envisioned a vibrant promo for a 'Weekend Special'—a pesto pizza paired with a complimentary drink. With just a few clicks, the AI would generate crisp visuals, a smooth voiceover, and dynamic animations, ready to upload to Instagram Stories.

Next, he delved into AI Images, an app boasting the ability to craft stunning graphics and advertisements. No longer restricted by his limited design skills, Javier was now free to experiment with infinite creative layouts, each image tailored and refined based on simple spoken instructions from his microphone. The concept was liberating; he could describe his vision audibly, and the AI would bring it to life within seconds. As he tested this feature, the scope of his marketing capabilities expanded exponentially before his eyes.

Excited, Javier moved on to Website-Building Apps and Funnel-Building Apps. The ease with which complex sales funnels could be established intrigued him. The app simplified the process of converting interest into sales, crafting a seamless journey where potential customers were guided through personalized touchpoints—from curiosity to checkout. By Sunday afternoon, he had mapped out funnels that could tailor communications to his diverse customer base, offering customized experiences that were as efficient as they were engaging. Another eye-opening feature was the possibility of integration with email and domain management apps. Here, Javier could see the utility in synchronizing email marketing with other aspects of customer outreach, ensuring campaigns were wide-reaching and profoundly personalized. Each email could be customized based on customer data pulled from integrated apps like CRM and chatbots, enhancing his ability to connect meaningfully with each patron.

A powerful idea struck him as he explored the Task & Process app. He could map out real-time production sequences, assigning responsibilities per station—tracking the number of pizzas, grams of cheese, tomatoes, and sauces like a streamlined production line. It wouldn't take away from the artistry of pizza-making, but it would introduce efficiency, reducing waste and optimizing costs. That would make it more efficient, enable the cost reduction that was needed, and make the buying activities more predictable. By using AI to modernize operations, Javier saw a path to preserving the family legacy while ensuring its future in a digital age.

As he learned, Javier felt his enthusiasm build with each passing hour. The landscape of digital tools unlocked potential that he had not fully appreciated before. Each app served as a new piece of the puzzle—AI-driven seminars fostering a deeper understanding of digital, automated engagement, and marketing mastery.

The weekend flowed in a blur of exploration and strategy mapping as Javier immersed himself in the digital toolkit at his fingertips. The more he learned, the more he saw his early steps into AI as transformative for his business's future—each app a new ally in the quest to sustain and scale his pizzeria in an ever-competitive market landscape.

By Sunday evening, a new sense of confidence filled him. Javier recognized that the tangible steps toward advancement lay in learning and applying these newfound skills to build strategies that would leverage AI for real-world benefits. The excitement of potential no longer felt like a distant chase but a close, achievable horizon—all thanks to the innovative blend of creativity and technology the platform offered him.

Unlocking Potential with AI

Seated at his desk, Javier was fixated on the video unfolding before him. It was part of the comprehensive AI Academy course that accompanied the MKTGAIPRO platform. Filled with a blend of anticipation and trepidation, Javier absorbed every word, his notebook filled with hastily scribbled ideas and potential questions to explore. He felt a bubbling excitement mingled with awe as he unlocked each layer of strategic insight the program offered.

In this module, the ever-charismatic Mike Rubence appeared onscreen, ready to impart another slice of wisdom to his virtual audience. Javier leaned closer, captivated by the promise of learning a novel technique that could redefine his approach to business growth. Today's topic was leveraging prompts in a language model to innovate and generate new product ideas—an aspect Javier had yet to master.

Mike began with an animated introduction, his passion shining through every word. "Welcome back, entrepreneurs! Today, we're diving into how you can use AI as a tool and a creative partner in your business journey. One of the most powerful aspects of working with a language model is crafting effective prompts and the precise, friendly conversation you might have with your AI."

After pausing to reinforce the importance of understanding this concept, Mike continued, "Think of a prompt as a conversation starter with your AI. It's not just a question—it's a precise cue that steers the AI toward meaningful, context-rich responses."

Javier watched closely as Mike illustrated the mechanics of prompting, explaining that an effective prompt is clear, concise, and sets the stage for the conversation. "When asking your AI for ideas, like developing a new product, you start with a specific goal in mind. Clearly state what you want the AI to consider."

Mike demonstrated this by typing, "Imagine we're creating a new pizza flavor that appeals to young professionals focused on health and sustainability. What ingredients and unique twists could we incorporate?"

"Notice," Mike explained, "how this prompt combines clarity with imagination. It suggests context while inviting the AI to innovate within specific boundaries. The specificity here helps the AI understand exactly what you want to achieve. Of course, learning to use these tools takes some time, but once you get the hang of it, the results will speak for themselves."

As the screen filled with the AI's suggestions—a menu of creative, innovative pizza ideas—Javier's excitement soared. The outputs were a testament to the AI's capacity to exceed expectations, conjuring delicious concepts that blended tradition with modern trends.

"You have everything you need to succeed," Mike urged with confidence, locking eyes with his audience through the lens. "Treat the AI like a friend—a collaborator who is ready to think outside the box with you. Let it surprise you with answers you might not have considered alone."

Javier nodded to himself, feeling the shackles of conventional thinking loosen. The barriers of self-doubt and creative blocks dissolved as he began to recognize AI as a tool and a dynamic resource ready to elevate his vision.

Closing the video module, Javier was filled with newfound clarity. The concept of prompts was no longer an abstract theory but a practical skill he could harness to breathe life into his evolving plans for the pizzeria. Embraced by inspiration and utility, Javier prepared to embark on this AI-guided journey, eager to unlock the untapped potential resting within both the technology and himself.

The video continued with Mike Rubence seamlessly transitioning into his next practical lesson—creating a powerful product brief using an AI-generated prompt. Now more attentive than ever, Javier felt a surge of excitement coupled with a flicker of anxiety about applying these concepts in real time.

Mike addressed the camera with his characteristic enthusiasm: "All right, let's solidify what we've learned with a practical example. We'll craft a prompt to develop a compelling product brief for a new pizza flavor. This isn't just about listing ingredients—it's about telling a story and envisioning how this pizza fits into your brand and customers' lives."

Javier leaned closer, eager to see how this theoretical knowledge could translate into tangible results. The idea of using AI to encapsulate his creative vision into a structured format was both exhilarating and slightly daunting.

Mike typed out his prompt methodically, explaining each component as he went. "Remember, a great prompt is detailed and directive, guiding your AI collaborator to flesh out an idea with depth and clarity."

Here's what Mike typed:

"Create a detailed product brief for a new, health-focused pizza flavor designed for young professionals who value sustainability. The pizza should feature unique, sustainable ingredients that appeal to health-conscious consumers. Include a catchy name for the pizza, a description of its flavor profile and ingredients, and marketing angles highlighting its ecological benefits and lifestyle appeal."

As the AI responded with an impressive array of ideas, Javier watched in awe, feeling a profound appreciation for the potential he now had at his fingertips. It was as if doors to a new realm of creativity were swinging wide open, inviting him to explore without fear of creative block.

Feeling inspired, Javier decided to try crafting his prompt using the structure Mike provided. The essence of effective communication through prompts made sense, and he felt ready to move beyond passive learning to active creation.

Taking a deep breath, he formulated his prompt:

"Generate a comprehensive product brief for a gourmet, globally inspired pizza. The pizza should combine exotic spices and flavors that intrigue adventurous eaters, appealing to foodies interested in international cuisine. Name the pizza, detail its flavor profile and ingredients, and suggest innovative marketing strategies highlighting its global inspirations and culinary creativity."

With this successful attempt behind him, Javier understood the mechanics of prompting and the boundless creative pathways it could unlock. The realization that he was fully embracing this new skill filled him with a gratifying sense of empowerment and renewed enthusiasm for what lay ahead.

As the concepts from the course settled in his mind, Javier couldn't help but draw parallels between mastering prompts and navigating the challenges in his own life—both requiring clarity, focus, and intentionality.

Javier thought about his health struggles, the constant ebb and flow of dealing with dialysis, and the strength he found in seeking guidance from Laura. Her coaching provided clarity and focus, a means of navigating the stormy seas of uncertainty that often swirled around him.

For Javier, the lesson on prompts crystallized a powerful idea: success lies in asking the right questions and providing clear direction, whether to AI or to himself.

Then his mind shifted to the transformation unfolding at the pizzeria with the aid of Mike Rubence and the AI platform. Just as Laura helped him manage his health, Mike was guiding him to harness technology, turning learning into action and potential into palpable success.

The parallel was undeniable—Laura helped him manage his health, just as Mike was guiding him in transforming his business. Both offered tools to simplify complexity, turning overwhelming obstacles into clear, manageable steps. In health, it was managing treatments and emotions; in business, it was integrating sophisticated technology into daily operations.

As these thoughts coalesced in Javier's mind, he realized that life, much like a puzzle, was beginning to take shape. Each element—health, business, personal growth—was a piece fitting into a larger narrative. The chaos he once felt was giving way to a vision where each component had its place, guided by knowledge, technology, and dedicated support.

He understood now that this journey was not about conquering challenges alone but about accepting help, integrating insights, and using them to build something greater than the sum of its parts. This was the puzzle he was solving, piecing together his strengths with the expertise and resources offered by those who crossed his path.

With this insight, Javier felt a serene acceptance of his path forward. He saw himself not just as a business owner or a patient but as a sophisticated strategist, equipped with the tools and insights to shape his narrative fully. He could see clearly how Laura and Mike were not just mentors but pivotal pieces that catalyzed his transformation.

As the night deepened, so did his resolve, the puzzle of life revealing a beautiful picture of interwoven strengths and possibilities. Javier embraced the moment, knowing that every step, guided by clarity and purpose, brought him closer to a future he could now envision with certainty and hope. Just as dialysis demanded discipline and trust in technology, so too did transforming the pizzeria—a journey of balancing resilience with innovation.

He saw his watch and glanced at his bed, then ran and threw himself onto it. He felt some pleasure in doing that. Then he took his diary to close his day and wrote: "Tonight, I feel a spark I haven't felt in years—the thrill of possibility. For the first time, the future doesn't feel like an uphill battle but a puzzle where each piece is finally clicking into place. Business and health, resilience and innovation, fear and action are all part of the same equation. The key is knowing how to ask the right questions, whether to AI or to myself. I close my eyes, knowing that the road ahead isn't just challenging—it's worth every step."

"As I reflected on Mike's lesson about prompts, I realized how much power lies in asking the right questions—not just to AI, but to myself."

CHAPTER 6
Understanding Your Audience

Early Sunday morning, light streamed through the pizzeria's windows as Javier settled into a booth, his temporary office. His untouched breakfast sat beside him, forgotten in his focus on the glowing laptop screen. Across the room, his mother tidied up from the night before, her presence a familiar comfort as he immersed himself in the AI platform he had recently invested in.

As Javier crunched through his research, he turned briefly to his mother with a warm smile. "Mom, can you handle the rest? I need to dive back into my AI strategy," Javier said, his tone appreciative.

She returned the smile, giving him a soft kiss on the cheek. "Of course, Javi. Just make sure you discover something wonderful," she replied, her eyes twinkling with the pride and support that always buoyed his spirits.

Her support was a balm, quieting the nerves that came with the weight of his new undertaking. Feeling invigorated, he turned back to his screen, where the AI platform awaited. Today's featured video starred Mike Rubence, as animated and engaging as ever, ready to share the secrets of audience engagement—a vital piece of Javier's emerging strategy.

The video commenced with Mike foregrounding the importance of truly understanding one's audience. "Today," he began, "I'm going to show you how a well-crafted prompt uncovers both the emotional and practical needs of your audience. By understanding what makes customers happy—or frustrated—you can create experiences that truly resonate, especially on tough days."

Javier watched intently as Mike navigated through the features, demonstrating how an LLM could weave insights from a well-structured prompt. "What we want to do," Mike emphasized, "is uncover the pain points and emotional landscapes of our target audience, so we can tailor our offerings to meet them where they are."

His fingers danced over the keyboard as he constructed a prompt designed to create a negative emotional diary for an audience aged 14 to 40—a tool to better understand their struggles and how his business might effectively address them.

"Think about the struggles and uncertainties people face," Mike continued. "Capturing these in a prompt gives you a map of their emotional terrain, helping you strategize your responses and interactions." Javier felt captivated by the elegance of the process, noting the simplicity yet profound depth of the technique. As Mike typed, responses from the AI revealed layers of worry and challenges faced by their audience, providing a roadmap of opportunity.

"Here's a straightforward prompt to tap into those insights," Mike stated as he shared the sample dialogue:

> "Write a diary entry after a difficult day. Describe the emotions, setbacks, and frustrations experienced by young people (ages 14-40). What specific challenges left them feeling overwhelmed or uncertain?"

Javier's eyes widened as he realized the power of this insight. The AI generated a wealth of scenarios that mirrored real-world struggles—unseen pressures, unmet aspirations, and the day-to-day grind that could leave anyone, anywhere, feeling less than whole.

"So," Mike concluded with a flourish, "knowing these pains isn't just about sympathy; it's about crafting avenues for empathy and solutions that resonate with your customers."

Feeling equipped with this newfound knowledge, Javier felt a burgeoning confidence as he crafted his prompt, eager to glean actionable insights from the AI. The answers tumbled in—a mosaic of daily struggles and emotional hurdles, each promising new openings for connection and engagement.

His prompt read:

> "Create an emotional diary entry for a young person aged 14 to 40, reflecting on a tough day. Explore the emotional and social challenges faced, their impacts, and how businesses like ours can provide solace or solutions."
>
> "For instance, a coffee shop used similar prompts to create a loyalty program that doubled customer retention by talking to their audience, addressing the pain points, and manufacturing special promotions to meet their needs."

Empowered by this exercise, Javier saw clearly how these insights would transform his approach, tailoring his business strategies to authentically address the genuine needs of his audience, creating not just products but experiences that truly mattered.

Alone in the quiet of the pizzeria, with only the hum of refrigerators for company, Javier felt the newfound thrill of discovery. Inspired by what he had achieved with the negative diary entries, he now sought to capture the flip side: the essence of a perfect customer experience. How did these moments shape a good day for his patrons? What would bring them joy and ensure a memorable visit?

Sitting back with a sense of purpose, Javier realized that understanding both ends of the emotional spectrum could give him a comprehensive view of his customers' experiences. Armed with this insight, he could not only address pain points but also amplify moments of joy, crafting experiences that left lasting impressions. While understanding challenges is essential, envisioning the joy he can bring to his audience allows him to craft truly memorable experiences.

With a few deep breaths to center himself, Javier began formulating a prompt that would tap into these positive experiences. He knew this required sensitivity and specificity, ensuring that the AI could generate vivid, actionable insights.

As he pondered what constituted a good day for his audience, his thoughts drifted to small pleasures: the interactions, the ambiance, the little details that turned a simple meal into an unforgettable occasion. He understood that the secret lay in the subtleties, those intangible sparkles that linger long after the last bite.

Determined, Javier crafted the prompt:

> "Write a diary entry after a perfect day. Describe the emotions and experiences of young people (ages 14-40), highlighting how a visit to our pizzeria made their day unforgettable. What interactions, services, or dishes brought them joy, comfort, or excitement?"

With the prompt sent through the AI, Javier held his breath, watching as words scrolled across the screen. The AI response was rich, painting a vivid picture of an ideal customer experience.

The description detailed a bustling pizzeria filled with laughter and warmth. Patrons talked animatedly, reveling in the comforting aroma of freshly baked dough mingling with the vibrant zest of tomato and basil. The atmosphere was inviting, enhanced by friendly staff members who greeted each guest by name, anticipating needs before they were voiced.

Customers spoke of the delight in discovering unique pizza flavors, the thrill of sharing meals with loved ones, and the comforting familiarity that greeted them upon entering. Energetic playlists curated by demographic preferences set the mood, while personalized menu suggestions delighted every palate.

Javier read through the AI's narrative, his heart lifting as he imagined his business embodying these qualities. The exercise crystallized his vision of the pizzeria as not merely a place to eat but a vibrant, welcoming hub where community and culinary adventure met seamlessly.

Javier realized that with these insights, he could refine his strategies to not only ensure satisfaction but also to make each visit to the pizzeria a cherished memory, ultimately living up to the family legacy with that modern twist he so ardently sought.

Navigating New Beginnings

Javier sat in his cozy office at the pizzeria, the heart of his world, surrounded by the familiar hum of morning routines. Sunlight poured through the window, casting a warm glow over the clutter of documents and the sleek gleam of his laptop. He was delving into the new world of possibilities that Mike Rubence's special one-on-one coaching programs promised. The allure of advancing his AI learning more swiftly than ever was irresistible. Here, in this familiar nook, Javier was on the verge of committing to yet another step in his journey to revolutionize his business.

"Just a few clicks," he thought, as he navigated the intricacies of Mike's offerings. His mind was alive with the idea of what such tailored guidance could unlock, the potential it held to leapfrog his understanding and application of AI in ways he hadn't yet imagined. His fingertips hovered over the buy button, a simple press away from potentially transformative insights.

His thoughts were interrupted by the entrance of Carlos, the head chef, a burly man with a genial smile that displayed both confidence and eagerness. "Javi, we have the new flavors of pizza you asked for," Carlos announced, his voice carrying a note of pride. "Come and taste."

Javier was jolted back to reality—a reality that was delightful yet fraught with a personal struggle. The chef's words invited a moment he had been looking forward to, yet it also begged the consideration of his health; a reminder of the doctor's strict admonitions about his diet echoed in his thoughts: less salt, low potassium, and phosphorus. A tightrope walk between business passion and personal health ensued.

"Right now..." Javier began, his voice trailing off as a cascade of emotions and thoughts vied for prominence. The anticipation of the tasting clashed with the stern reprimands from his last medical visit, bringing a tightening in his chest.

As he left his chair and moved to the door, the significance of what to convey to his team weighed on him. His role as leader required openness and honesty, grounded in the bedrock of personal responsibility. He knew that his credibility, both as a leader and a team player, was staked as much on honesty about his constraints as on the vibrancy of new ideas.

Javier hesitated, then met Carlos's gaze with a steady smile. "Carlos, let's check them out. But this time, I'm trusting your expertise. You'll lead the tasting—I need your palate now more than ever." His words steadied him, a gentle nod to his condition without overshadowing the enthusiasm that Carlos and the team were about to unveil. This acknowledgment turned his worry into an ally, focusing more on strategic delegation than immediate pleasure—a facet of the leadership ripple effect he aimed to instill.

Walking into the bustling kitchen, Javier was engulfed by the vibrant clatter and savory scents swirling thickly under the high ceiling, each carrying with it the essence of shared labor and creativity.

A new lesson was revealed to Javier: a realization that unfolded as part of an internal strategy. Transparency wasn't solely about exposing fragility but about harnessing it with intelligence; it was about enabling his team members to rise boldly in roles he'd support for the good of all. In tasting metaphorical and literal flavors in life, there were always choices that fulfilled more than immediate craving—choices that left an enduring taste of responsible growth, a nourishment.

As Javier navigated the tasting session, his thoughts kept circling back to Mike's upcoming program. He mused on how this represented another layer of enrichment, akin in some ways to Carlos's investment in the culinary creations taking shape. The parallels between savored bites and the satisfaction of seeing an idea brought to fruition intrigued him. With each taste, Javier confirmed the potential in both his team's talents and the promising avenues technology could bestow upon tradition.

Seated back in his office, flavors and insights still palpable, Javier returned to the glow of his computer. His decision settled firmly, he finalized his purchase of the coaching program—not just in search of efficiency and success, but for the sharing of an experience that promised to broaden horizons, much like Carlos's new pizza flavors hoped to do for patrons venturing to reach for more on their plates.

In the reflective solitude that followed the morning's activities, Javier marveled at how his earlier trepidation had transformed into measured confidence, understanding how decisions, whether about diet, business, or new avenues of learning, all required the same blend of caution tempered with fearless exploration.

As Javier leaned back, a sense of synchrony between his physical limitations, business ambitions, and a shared vision with his team swelled within him. He recognized that connections often emerged unexpectedly, like notes in a melody that, once begun, structured a harmony. With a fresh perspective settled deeply into place, the path forward offered not just challenges but a vibrant tapestry of collaboration and discovery reflective of a day that began with the simple comfort of a booth and a mother's steadfast encouragement.

Javier stood at the center of the pizzeria's bustling kitchen, the familiar scent of herbs and spices mingling with the warmth of expectation. His team gathered around him, forming a close-knit circle infused with years of shared work and camaraderie. Each face reflected curiosity mixed with reverent attention, knowing that momentous words were about to be shared.

Taking a deep breath, Javier began, his voice steady yet infused with sincerity. "Thank you all for joining me today. As you know, we are facing times that demand both resilience and innovation. It's about embracing change to secure a thriving future for our beloved pizzeria."

The team listened intently, the gravity of his words resonating deeply. Javier paused for a moment before addressing his personal challenges, aware of the vulnerability it would require. "I'm sure some of you know that I've been managing some health issues. Starting dialysis three times a week isn't just a routine adjustment; it's a new reality for me. Alongside this is the commitment to a disciplined diet that now dictates my role here."

As his words settled, whispers of support threaded through the room, eyes meeting with understanding and empathy. Javier drew strength from their concern and continued, "This leads me to the heart of our conversation today—change. I won't be stepping into the kitchen for tastings anymore. It's a bittersweet transition, but I am confident in leaving this responsibility with Carlos and every one of you, as you are the lifeblood of this kitchen."

Carlos, visibly moved, exchanged a brief nod conveying his readiness and gratitude for the responsibility bestowed upon him.

"But with this shift," Javier continued, his eyes brightening with determination, "comes an exciting new focus. I will pour my energy into crafting innovative ideas into reality and driving our digital marketing strategies. Together, we will design a plan that ensures our pizzeria is not only vibrant but packed full of patrons appreciating everything we offer."

The team felt energized by his words, their unease transforming into an eager anticipation of this promising direction. The unity of intent was palpable, a shared breath of purpose animating the room.

"I want you to remember," Javier's voice grew stronger, more resolute, "we are a family beyond business. Many of you have worked alongside my mother and have been with us through thick and thin. But as we move forward, we embrace a new style—one that blends tradition with a fresh perspective. My trust in myself extends to all of you. Together, let's pack this place up, brimming with success and shared pride."

Applause erupted spontaneously, a tapestry of claps and cheers that enveloped Javier in a warm embrace of approval and solidarity. Carlos stepped forward, eyes glinting with pride and emotion.

“How about a ‘no-salt special’ in your honor, Javi?” he said, drawing laughter from the team.

Then, with great leadership potential, he embraced Javier with a hug and said, "Javier," he began, his voice laced with emotion, "we stand behind you. Your faith in us ignites a fire, a drive to ensure your vision becomes a reality. Your courage in sharing your journey strengthens our resolve. We won't just support you; we'll walk with you."

As Carlos finished, the team's affirmation solidified a powerful bond, forging a collective resolve to meet the future with unwavering enthusiasm and dedication.

Javier returned to his office, the echoes of their support still resonating vividly within him. He sank into his chair, reflecting on the depth of Carlos's response. A profound sense of validation and purpose enveloped him. The team's reception to his announcement had reignited his confidence, relieving a weight he hadn't fully acknowledged until clarity swept over him like a refreshing breeze.

Javier's heart swelled with gratitude for the team that shared his vision, moved by how their reaction transcended words—a tangible support that embraced his aspirations and health with equal passion. Here was a microcosm of possibility where shared courage and mutual trust promised not just survival but flourishing and harmony.

With renewed determination, Javier envisioned the path ahead, one imbued with innovation, creativity, and the unfaltering spirit of a team united in purpose. The journey was now a shared one, rich with challenges yet equally generous with the potential for uncharted success.

Javier sat quietly, the soft glow of his laptop screen illuminating the dimly lit room. The Zoom meeting with Mike Rubence had concluded moments ago, yet the echoes of their discourse lingered vividly, a harmony of new knowledge and encouragement that played over in his mind.

The call had been invigorating, and Javier felt a lightness in his chest—an unfamiliar yet welcoming sensation of clarity and inspiration. Mike had been a great mentor; his guidance was precise and profound, and he effortlessly wove wisdom with practicality.

"The great power of AI is not in itself," Mike had emphasized during their conversation, "but in the fusion of your personal experiences empowered and increased by AI tools. So be confident in your experience and enjoy the path of learning how to increase and empower your experience."

In the quiet aftermath of their chat, Javier found himself reflecting deeply on these words. There was a truth to them that felt resoundingly personal, resonating with the core of his endeavors and trials. The idea that his unique experiences, both personal and professional, were not mere hurdles but rather intrinsic assets laced with potential untapped strength was a revelation. It brought him serendipitous freedom and a license to view AI not as an external force but as an extension and amplifier of his vision. "Just as I shared my struggles with my team, I want our customers to feel that we truly understand their stories."

Yet beneath that comforting revelation, a whisper of uncertainty threaded through. Could he truly rely on his personal experiences as a foundation for such ambitious aspirations? Doubts tugged at the edges of his confidence. The task of creating a lead magnet—a foundational strategy designed to attract and retain customers—loomed large, inviting self-reflection and scrutiny.

Embracing a role so predicated on self-referential clarity was daunting, a challenge demanding faith in his narrative and decisions. But what Javier held in his grasp now, thanks to Mike's insights, was the understanding of the symbiotic relationship between his life's journey and the technology he was now wielding. He realized that the fusion of past experiences and emerging tools held transformative power—the kind that required not just ambition but thoughtful execution.

The juxtaposition of fear and potential settled warmly within him, a reminder of the adventures awaiting just beyond the threshold of today. He recognized that stepping into this new endeavor was akin to setting a fresh melody to the overarching symphony of leadership and growth he was composing.

Javier stood by the window, gazing into the night, kissed by the gentle glow of streetlights, ready to embark on the task of crafting his lead magnet. The path forward would undoubtedly twist with challenges and insights, but an exhilarating journey had never felt more deserving of pursuit.

Buoyed by the evening's affirmations, Javier finished the scene prepared, if still hesitant. The rumblings of self-doubt were no more than fading echoes as he set thoughts to action, lit by the promise that his authentic self, empowered by AI, could indeed chart the vibrant course he so fervently imagined.

Investing in Self

As the clock inched towards eight, Javier sat in the familiar waiting room of Laura's consultation office. The calming hues of the early evening sky filtered softly through the windows, casting a serene ambiance that countered the swirling currents of thought within him. Today felt different—fuller, yet laden with a sense of cautious introspection.

"Javier?" Laura called, her voice a familiar herald of safety and introspection. Rising, he stepped towards her office, each stride heavy yet driven by the anticipation of release and clarity.

Inside, the room offered warmth; framed certificates silently bore witness to their upcoming conversation. Laura seated herself with her characteristic poise, her eyes reflecting understanding and patience. "So, how are you today?" she began, her voice inviting revelation.

Javier took a deep breath, diving into an account seasoned with pride and lingering doubt. "I talked about my health challenges with the team, lauded their incredible efforts, and shared the exciting developments of joining Mike's coaching program. We've set our sights on launching a new menu," he said, weaving a narrative that showcased both external achievements and internal evolutions. "I feel the changes I'm making personally, but there's this nagging sense. It feels... too self-referential."

Laura nodded, digesting his words with her natural rhythm of consideration. "How would it have to be if it wasn't about you? And how would people find it authentic and special?"

Silence settled between them, filled with the low hum of introspection. Javier's mind danced through narratives of influence, diverging along paths of personal growth and relational authenticity. His gaze drifted, anchoring onto an abstract painting adorning Laura's wall, colorful swirls reflecting the fluidity of his thoughts.

How indeed, if not about himself? What was the bridge between personal introspection and genuine communal benefit?

Laura's voice softly broke through his reverie. "What do you know better—yourself and your customers, your community, or an external pizza menu?"

"Myself and my clients," Javier replied, the realization threading clarity through his earlier doubts. An evocative understanding unfurled within him, rippling outward, transforming unsettled knowing into grounded conviction.

"So, you know what you need to do. Gamble all in on you this time. Have you done that?" Laura challenged; her words were gentle yet penetrating.

Javier felt the surge of new confidence, akin to witnessing the dawn over an internal landscape suddenly seen in sharper relief. The pieces aligned, building towards a comprehension that what truly held transformative power was the authenticity he shared with his patrons—understanding their stories and preferences because they mirrored his own journey and aspirations.

As their session ended, Javier left Laura's office draped in a newfound sense of direction. The evening air embraced him as he began his walk home. The streets, bathed in the golden glow of streetlamps, pulsed with life—couples strolling, children at play, an animated bustle that spoke of community and shared spaces.

Each step resonated with the promise of renewal, whispering possibilities anchored in a profound reimagining of his role—a custodian of stories; an architect of communal experiences that transcended individual success. He reflected on the coaching wisdom imparted by Mike, now laced with the depth of Laura's sagacity, transforming abstract lessons into avenues of thoughtful application.

By the time Javier reached his doorstep, a tapestry of intentions woven tightly around belief in the impending morning, he knew where he needed to go. Tomorrow's path was not just about turning ideas into products but about transforming stories into symphonies—a celebration of personal, shared, and culinary tapestry.

As he closed the door behind him, hopeful blends of purpose and authenticity illuminated the way forward in a vibrant crescendo, promising a harmonious future invitingly within reach.

Javier shuffled into the warmth of his home, the day's journey imbued with newfound understanding. As he entered the kitchen, the familiar surroundings encircled him with comfort. He moved with purpose toward the coffee maker, the ritual of brewing a cup a soothing balm for the senses. The rich aroma filled the air as he settled at the kitchen table, the steaming mug nestled in his hands.

His eyes drifted to the diary resting nearby, its cover worn yet inviting with the potential to encapsulate another chapter of growth and discovery. The pen felt right in his hand, a beacon guiding him through thoughts yearning for expression.

He began to write, each word a reflection of the day's culmination and the lessons enshrined within it:

> "Tonight, as I sip my coffee, clarity settles over me. Today unfolded like a tapestry—each conversation, each realization weaving into a greater picture of growth.
>
> Speaking with Laura made me realize the essence of self-investment. I've long feared that my endeavors appeared too self-centered, yet the dialogue unraveled that misconception. It is not self-centered to draw from one's life experiences; it is the most authentic form of creating something meaningful. What I bring into this world—be it

into my business or my connections—is my truest offering, drawn from the depth of my understanding of myself and my community."

Mike reminded me that AI isn't just a tool; it's a force multiplier. My experiences, struggles, and insights shape how I wield it. The true power isn't in technology alone—it's in blending it with human connection, turning business into something deeply personal. Understanding both struggle and joy is what makes a place feel like home. And that's exactly what I want my pizzeria to be.

As I reflect on the meeting with the team, I see how this understanding extends beyond personal growth—it seeps into the foundation of trust and collaboration we are building. When vulnerabilities are shared openly, they transform into pillars of collective strength.

The streets I walked tonight mirrored this freedom, bustling with life, reflecting a world rich with interwoven stories, possibilities echoing through every interaction. I know now that tomorrow holds the promise of strategies that will ripple from the honesty of my experiences to the patrons we serve, turning our pizzeria not just into a business, but into a living conversation.

As I close the pages of this diary entry, the lessons settle within me, not as mere thoughts but as actions that guide my hand as I move forward. Here's to another day of integrating authenticity and innovation—a journey as boundless and rich as the simple pleasure of sharing a meal with those I cherish.

Today felt like turning a key in a lock, opening a door to possibilities I hadn't fully realized before.

I pictured a bustling pizzeria where every customer felt seen and valued, their stories woven into the fabric of our success.

Javier set down the pen, a gentle sigh of contentment escaping him. He rose from the table, the day having folded into a box of insights and dreams, neatly tucked away for tomorrow's unfolding. With hope and determination intertwined, he turned off the kitchen light and carried the rhythms of the day into a restful night.

CHAPTER 7
Crafting Knowledge and Creativity

Morning light spilled across Javier's workspace, painting his desk in soft gold as he settled in at 8:00 AM. Today was the day to turn knowledge into action—to craft a lead magnet that would engage and inspire his audience: an e-book that blended creativity with purpose.

Inspired by Mike's video tutorial, Javier felt a surge of purpose. His idea had a unique twist—an e-book of pizza recipes crafted for those managing kidney conditions, proving that great taste and wellness could go hand in hand. This concept would showcase an intersection of culinary delight and wellness—an acknowledgment of his own health journey and a gift to those in similar situations.

Javier eagerly opened the Book Daddy app on the MKTGAIPRO platform, anticipation tingling at his fingertips. The app greeted him with a streamlined interface designed for intuitive creativity. As he delved into the first step, character creation, he marveled at the ease with which personalities emerged. Each character carried a unique story, adding depth and resonance to the narrative nature of his e-book.

Next, the app's intelligent outline generator kicked in, sketching a structure Javier could tweak to perfection. As the blueprint unfolded on the screen, he marveled—it felt like watching an artist bring an idea to life, each section appearing like brushstrokes on a canvas. Javier tuned the outline to align precisely with his vision, selecting a title that captured the heart of his concept while drafting chapter-by-chapter content, leaving him both impressed and excited.

With the blueprint of his e-book laid out, Javier transitioned seamlessly into the Ghostwriter feature—a tool that promised to bring his narrative to life. This section of the app invited him to describe what unfolded in each chapter, guiding AI-generated prose with his input. Javier detailed scenes with vivid descriptions and specified the tone he wanted, eagerly watching as each scene was crafted before his eyes.

The progression from concept to creation was surprisingly fluid, and the efficiency with which this tool operated felt astounding. As each scene unfolded, Javier's delight grew; the AI's ability to embody his imaginative threads and express them in story format was both thrilling and reassuring. Even as he worked through the day, building the draft chapter by chapter, a childlike excitement threaded through his dedication—an intangible blend of creativity and technology manifesting into reality. As he typed the final prompts towards the end of the day, anticipation swelled to a crescendo. The culmination of snippets, scenes, and synthesized knowledge lay before him—a full draft of his eBook. Javier sat back in his chair, a sense of achievement flooding him. What had begun as an idea was refined into an actualized working draft, breathing with character and substance.

The day had been immersive, a journey showcasing modern tools empowering creative aspirations. Understanding this, Javier realized the path was not only about the project itself but about learning to leverage his unique narrative in a way that embraced both innovation and authenticity.

With the first draft complete, he felt invigorated, inspired by what was possible with the right tools and a fully engaged mind. Tomorrow would welcome refinement, but today celebrated creation—the harnessing of personal and digital ingenuity to offer value and connection through every page crafted.

Javier sat at his desk; his laptop screen alive with the interactive platform of his coaching session with Mike Rubence. The momentum of their discussion brimmed with insight, underscoring the intricate yet invigorating journey they had embarked upon together. As the session neared its conclusion, Javier listened intently to Mike's suave, directive voice—a mentor weaving a tapestry of strategy with confident ease.

"So, Javier," Mike began, leaning slightly forward, his demeanor radiating encouragement and clarity, "you've done fantastically well using the AI platform with its GPT capabilities to coordinate the initial stages of your concept."

Javier nodded, his mind reflecting on the phases they had traversed: crafting a product brief, building a vivid avatar or buyer persona through diary entries exploring both challenging and successful days, and most recently, the development of his lead magnet—an eBook brimming with fresh, health-conscious pizza recipes.

"Next up," Mike said, leaning in with a grin, "we're tackling your Video Sales Letter. This is your pitch—your chance to hook people instantly. I'd suggest collaborating with someone from the pizzeria to keep it real and authentic."

Mike's guidance unfolded a clear roadmap, layered with intentionality. He elaborated on each step with careful precision, ensuring Javier grasped the nuances: "Once your VSL is ready, we'll pivot to leveraging other AI tools you've mastered. Start with the AI videos to visually enhance your promotion, bringing dynamic animations and voiceovers to life."

Javier absorbed the practical applications, envisioning the fluidity with which he could transition ideas from concept to vivid digital presence.

"AI Images," Mike continued, "will provide you with captivating graphics, tailoring every visual aspect to resonate with your created Buyer Persona. And then, Ultra-Fast Funnels Flex—our secret component, a tool tailored to streamline and optimize your promotional funnel to maximize exposure and engagement."

Through these phases, Mike crafted a vision that framed not just the process but the philosophy of blending AI with entrepreneurial ingenuity—integrating technology in a way that elevated both reach and authenticity.

Javier felt a burgeoning surge of motivation with every word, each step outlined by Mike framing a ladder of accessible goals. The synergy between his deepening understanding of AI's capacity and personal creative intent stoked a vibrant determination within him. Here was a scaffold that, once navigated, promised the manifestation of his trajectory from aspiration to practical success.

The coaching call ended with Javier suffused with gratitude and anticipation. Mike's comprehensive guidance ignited a balanced mix of readiness and the exciting unknown—a space where calculated effort meets destined realization.

As the call ended, Javier was left alone in the gentle dusk of his room, the laptop's glow a comforting companion. The checklist formed during the call lingered in his mind, each point a note in a melody soon to be orchestrated. With plans ignited and strategies charted, the path unwound invitingly before him—embracing a methodology that blended the essence of personal narratives with the resonance of digitally redefined realities.

Feeling a renewed sense of empowerment, Javier logged into the AI platform and navigated to the GPT interface. The clarity and structure of Mike's teaching resonated as he carefully composed his prompt for the Video Sales Letter (VSL). His focus honed, the prompt emerged with intent, a distillation of his vision for the eBook:

> You are an expert in creating video sales letter scripts. I want to create a Video Sales Letter script that will introduce this book to pizza lovers, offering them a manual or resource guide to learn the steps and take the actions to have an opportunity more than they ever thought possible. Please make this a captivating video sales letter script about 90 seconds long. I want it to be in the first person from

Javier, formatted to play directly in the teleprompter with no cues or background.

Javier hit Enter. Within seconds, words began spilling onto the screen—smooth, precise, captivating. He watched in awe as the AI crafted his script with a speed that felt almost surreal.

Script Generated by GPT

"Hey there, fellow pizza lovers! I'm Javier, and I'm thrilled to introduce you to something truly special—an e-book that's not just another pizza recipe collection but a gateway to delicious discovery.

In these pages, we unlock the secrets to crafting pizzas that aren't only tasty but also health-conscious, perfect even if you're navigating dietary needs like kidney modifications. Imagine diving into authentic recipes that elevate your homemade pizza game, filled with tips and techniques you won't find anywhere else.

Whether you're a culinary novice or an experienced home chef, this guide is packed with opportunities to transform your kitchen adventures into flavorful successes. Let's embark on this journey together—your personal guide to mastering pizzas that excite and inspire every single day.

Grab your copy today, and let's make pizza magic happen!"

As the script materialized on the screen, Javier was captivated. The succinct, engaging language captured his vision and message perfectly—a fusion of passion and promise delivered in a compelling format that breathed life into his project.

Amazed by the process and the output, he felt the culmination of his learning crystallize into something tangible, ready to share with the world. The efficiency and dynamism of the AI in crafting a message that resonated so deeply left him in awe. What started as an idea now had the potential to captivate and connect with his audience, poised to be further brought to life through the lens of a video.

With a satisfied smile, Javier appreciated the journey he'd undergone—the journey of creating, learning, and crafting a powerful tool to engage pizza enthusiasts—a step forward in both his personal growth and business evolution.

Navigating the Next Chapter

Javier sat at his kitchen table, cradling his phone as he dialed the number for Jonathan, the book editor he'd been meaning to consult. Anticipation thrummed through him alongside a host of questions swirling in his mind—questions born from the uncertainty of entering the publishing realm for the first time.

"Hello, Jonathan speaking," came the amiable voice on the other end, grounding Javier with its familiarity.

"Hi, Jonathan, it's Javier," he began. "I'm really excited about the draft of my new eBook, but I need your expert eye on it. Could you read it through and suggest any changes? I also need guidance on the steps to get it published."

Jonathan responded with warmth, seasoned with the candor of someone familiar with nurturing new authors. "Of course, Javier. I'd be happy to help. First, focus on making sure the structure flows naturally and that any technical details are polished. I'll go through it thoroughly and provide feedback, as we want it to resonate authentically with your audience."

Javier's appreciation flowed with his next question, "That's perfect, thank you. And what's the process to publish it?"

"Great question. Once we're happy with the draft," Jonathan explained, "we investigate formatting the eBook for various platforms—Kindle Direct Publishing is popular for self-publishing. You'll need a catchy cover, which I can assist with, and a compelling book description. Then we explore uploading it, handling metadata, and selecting categories to boost discoverability."

Javier took notes, processing each step as a tangible block in his structured journey. "It seems straightforward, but can I rely on help for these elements to ensure quality?"

"Absolutely," Jonathan reassured. He explained how metadata, like keywords and categories, would help readers discover the eBook. "Think of it as the GPS that leads them to your book," he said. "We can coordinate with professionals for cover design and formatting if needed. Marketing is crucial too; involving social media and leveraging that VSL you mentioned could drive interest."

Javier considered the scope of the path ahead, a blend of creativity and tactical execution. His concerns, while softened, remained about the practicalities of each stage.

"Thanks, Jonathan. This eases my nerves a bit."

"No worries, Javier. It's about finessing what you have into something cohesive. You've already done the hardest part by drafting your ideas. Publishing will magnify that effort."

Ending the call, Javier felt a complex mesh of excitement and apprehension. The clearer understanding provided by Jonathan calmed his immediate fears, presenting a path from creation to publication.

Sitting back, he pondered how this new knowledge dovetailed with Mike's guidance and the tools at his disposal. His earlier anxieties about the uncertainties of publishing had shifted down to a balanced view of opportunity bolstered by strategic support and expert insight.

The afternoon sun cast long shadows across the pizzeria's floor, its warm glow reflecting Javier's rising anticipation and the quiet buzz of potential filling his mind. As he placed the phone receiver back on its cradle, the satisfying click echoed softly in the room. He leaned back, running a thoughtful hand through his hair, steadfastly determined to turn plans into action.

The call with Jonathan, the book editor, had ignited a renewed sense of clarity and focus. Javier's mind was a tapestry of strategies and possibilities, each thread intricately woven with the promise of innovation and success. The symphony of upcoming actions played rhythmically in his head as he prepared to execute his next strategic move.

"Alan," Javier called out, beckoning the administrative executive into his office. Alan's approach was accompanied by the gentle rustle of papers, a signifier of dedication and readiness etched into his every movement. Stepping into the office, Alan cast an inquisitive glance, ready to absorb what Javier's mind had brilliantly concocted.

"Alan, I need a comprehensive report," Javier said, his voice carrying the weight of renewed vigor. "Please identify for me the days and times when our foot traffic dips below ten customers. We've got to tackle these weak spots strategically."

Javier's request was a blend of precision and urgency, each word accentuating his commitment to uncovering and addressing operational inefficiencies. Alan's attentive nod mirrored Javier's resolute determination, the mutual understanding igniting a sense of camaraderie between leader and supporter.

"Understood, Javier," Alan replied with certainty, the sound of his voice imbued with shared mission and resolve. Stepping back into the corridor, Alan allowed the office door to close seamlessly—a parting gesture that encapsulated a promise to deliver.

With Alan gone, the office fell into a gentle silence, the only noise a distant murmur of the kitchen preparing for the evening's patrons. Javier leaned forward, propping his elbows on the desk as his hands clasped together—a thoughtful stance interrupting the mental ripple of tactics and aspirations.

Javier revisited Mike's upselling sequence mentally, savoring the intricacies of engagement that moved stepwise from an accessible eBook to more intimate, high-value experiences at the pizzeria. Each stage of the plan was artfully stitched together into a cohesive strategy, a seamless rise in engagement and value that Javier aspired to execute flawlessly.

Lead Magnet eBook - US$2: A taste of knowledge, drawing enthusiasts into their flavorful oasis.

Practical Video Classes - US$150: Accessible mastery paired with hands-on learning.

Live Course at the Pizzeria - US$450: A gathering, grand in experience and enriched by in-person expertise.

Personal Coaching at Home - US$3000: An experience, luxurious and intimate, redefining the boundaries of culinary art.

The eBook acts as an entry point, sparking curiosity and establishing trust. The video classes deepen engagement, offering hands-on learning. By the time customers reach the in-person course or personal coaching, they feel fully immersed in the experience.

As Javier wrapped up his work on the eBook and an ascension product ecosystem, he realized that creativity wasn't limited to the digital tools he had mastered—it thrived in the quiet moments where personal experiences and simple joys intersected.

During his contemplation, a pause punctuated the air—a lull disrupted by the gentle rise of a familiar melody weaving its way into the office. It caressed the edges of silence with affectionate nostalgia.

Dean Martin's "That's Amore" drifted through the air, a delicate tapestry of sound inspiring a moment's pause. With it, Javier's surroundings assumed a whimsical hue, transforming the mundane into a narrative ripe with potential.

As the song reached his ears, the world outside his window softened, allowing space for the beauty of the melody to capture his mind. The music breathed life into the uncharted corners of his imagination, summoning forth a scene vivid with inspiration.

Javier felt its warmth seep into him—a comforting embrace akin to a heartfelt hug. The melody expanded, wrapping around his ideas and bringing forth a symphony of creativity and clarity that resonated with harmonious precision.

He recalled evenings at the pizzeria when couples lingered over shared slices, their laughter echoing warmly. Those moments of connection had always felt magical—now, he could recreate that magic in a new, intentional way.

In a brilliant flash of insight, the harmony between pizza and love declared itself. Pizza and Love—the essence of Dean Martin's gentle poetry and joyful celebration made manifest in a concept that harmonized perfectly with the pizzeria's spirit.

Javier envisioned laughter without restraint, flour floating through the air like snow on a gentle breeze, and the joyful chaos where the kitchen transformed into an Eden—a place where culinary adventure intertwined with tender connection. Partners and friends embraced the joyful chaos, reveling in shared experiences.

He imagined the gentle kneading of dough, hands synchronized with playful enthusiasm—a symphony of flour-dusted gestures mirroring the delectable cadence of "That's Amore." Couples crafting their pizza creations and molding them into shapes that echoed their shared journey—could anything embody love more aptly?

The idea struck like a perfectly timed note in a love song. His heart pounded in sync with the melody, each beat echoing the clinking of glasses, the laughter of diners, the warmth of a shared meal. "Amore Pizza," he whispered, the name rolling off his tongue like a lyric waiting to be sung.

This concept had legs—a grand idea conceived from simple inspiration, thriving on the nourishment of personal experience and rich ambiance. Such unassuming delight refined into an immersive culinary experience would bring a fresh melody to the beating heart of his pizzeria—a reframing of friendship and flavor intertwined.

Reflecting on the notion that profound ideas spring forth from the essence of the familiar, Javier embraced the essence of his realization—a recognition that music possesses the power to unlock creativity and amplify the familiar world with unparalleled beauty.

"The answer," he whispered, feeling a tingle of awe radiate through him, "has been here all along, in the notes of a favorite melody and the simple joy of togetherness."

Armed with revelation and brimming with anticipation, Javier committed to turning inspiration into reality—an embodiment of "That's Amore" distilled into an experience as rich and authentic as the song's source. His vision now firmly entrenched in shared laughter and culinary discovery—the lyrics of life played in perfect harmony with melody's muse.

As the night draped a quiet stillness over Javier's room, he lay in bed, chased by the intoxicating dazzle of inspiration that had gripped him throughout the day. The concept of the course—a romantic culinary escapade—played softly against the backdrop of his mind's eye, the evening's discoveries threading effortlessly into potential and promise.

Suddenly, a name crystallized with clarity—Amore Pizza. The name evoked a charm so naturally aligned with the very essence of what he wished to create: a romantic course centered around fun, where laughter mingled with the crafting of flavors. It was an invitation to both couples and single adventurers seeking connections through the shared joy of culinary exploration.

This vibrant idea felt like dawn breaking, washing over him with the gentle warmth of satisfaction and anticipation alike. Awakening his readiness to inscribe this newfound clarity, Javier reached for his diary, the pages whispering under his touch as they unfolded to absorb his thoughts once more.

Tonight, as I settle into the quiet of reflection, I find myself marveling at the concept that unfolded today: Amore Pizza—a romantic culinary escape that captures the essence of shared moments and connection. This idea did not just arrive; it flowed, nurtured by music, creativity, and the culmination of efforts threading through the fabric of my experience on a day-to-day basis. I marvel at the way inspiration finds us—in a melody, a memory, a moment of quiet reflection. The idea for Amore Pizza feels like more than a concept; it's a celebration of connection and creativity, a reminder that even the simplest joys can transform into something extraordinary.

Tonight, these pages don't just capture ideas—they weave a map of discovery. Every insight, every note, is a step toward something greater. The way creativity and strategy intertwine is nothing short of transformative. Alan's report directive, Mike's upselling path, and the delightful harmony of inspiration stringing it together—each has played a role.

I've realized the importance of observing beyond the surface and letting simple elements, like a favorite song, illuminate the endless possibilities of innovation. Creativity is not bound by complexity; often, what resonates deeply is found in the texture of the familiar, the richness of cherished ambiance.

My lesson is clear: The boundaries that once felt limiting are breaking down. Therein lies an awakening—a renewed ability to see not just problems but intricate canvases upon which solutions emerge, beckoning action. My path now revels in undiscovered avenues blending creativity and economy, as we weave connections both culinary and communal.

As I put pen to paper, I internalize this: Integrating life's melodies and experiences into business does more than simplify—it elevates. The possibilities mirrored in the course await; they are not merely aspirations but the harmonics of progress, inviting us onward.

I close this entry with the glow of unapologetic anticipation—a forthcoming journey in which ideas will dance with reality, each step challenging me to amplify what is real, inspired, and ever deeply human.

Javier exhaled a gentle sigh of contentment, closing the journal. What enveloped him now was a peace lined with the buoyancy of intention and understanding, offering not only today's digest of ideas but also tomorrow's promises as he drifted into dreams alive with potential.

CHAPTER 8
Fear into Fuel

Morning sunlight poured through the windows, casting a golden glow over Javier's office. Today felt different—charged with purpose. As his computer hummed to life, he logged onto the MKTGAIPRO platform, ready to turn anticipation into action. Today was pivotal; it was a day to channel creativity into concrete action—a launchpad for new beginnings through a meticulously crafted funnel.

In the solitary embrace of his office, Javier's fingers danced across the keyboard, configuring the intricate stages of the funnel designed to captivate and engage. The journey began with a small yet impactful video, meant to capture the vision and excitement of both his new eBook and the much-anticipated Amore Pizza course. The interface before him was alive with options, each click an affirmation of his readiness to reveal what he'd built.

With a deep breath, Javier reviewed the words he would share—words intended to resonate, to invite those who shared his love for pizza and community into the experience he'd so carefully crafted. Clicking the camera icon, he began the recording, his reflection staring back at him.

Video Dialogue

"Hey there, pizza lovers! I'm Javier, and I can't wait to take you on a journey where flavor meets experience.

Now, I know what you're thinking—'Another pizza recipe?' But this isn't just any recipe guide. It's a curated collection with a twist—combining health-conscious ingredients that are perfect even for those with dietary considerations. Discover how you can enrich your everyday meals with flavors that burst with zest and goodness.

And for those craving more than just an eBook, I'm thrilled to announce 'Amore Pizza'—a one-of-a-kind course at our pizzeria that invites couples and singles alike to craft, connect, and, most importantly, enjoy. This is more than cooking; it's about creating memories!

So why wait? Dive into the world of flavor and fun with our special offer available for a limited time.

Grab your copy today, and let's make culinary magic together!"

As Javier recorded his video, he couldn't shake the thought that this project symbolized more than just growth for the pizzeria—it was his commitment to living fully, despite the challenges ahead. The recording light dimmed as he finalized his video, a satisfied smile breaking across his face. Javier leaned back, his mind afire with excitement and the fulfillment of blending passions with purpose. Analyzing the funnel, he meticulously reviewed every stage—each interaction designed to lead seamlessly into the next, a testament to both design ingenuity and nuanced strategy.

Yet, as satisfaction settled in, something shifted—an unfamiliar weakness creeping into his legs, subtle at first but growing insistent. The keys beneath his fingers blurred. His breath shallowed. What was this? His legs, strong and reliable, suddenly felt unsteady, a slight tremor disrupting their usual surety.

The tremor spread subtly, a quiet alarm beneath the surface of his skin. Javier attempted to stand, driven by an unconscious need to distance himself from this uncomfortable sensation, but his legs seemed reluctant to obey. It wasn't painful, nor was it dramatic, but it was undeniably there—a disquieting manifestation whose presence kindled an unfamiliar concern.

His mind raced back through the collage of recent days, trying to grasp any preceding hint of such symptoms, but there was none. The novelty of the sensation startled him, and he couldn't help but dwell on its implications.

Could this be linked to his ongoing health challenges? Was it a solitary occurrence, or did it foreshadow something deeper? The uncertainty began to weigh upon him, a persistent whisper threading anxiety through his resolve.

He pondered his regimen, recalling Dr. Ruiz's cautionary words about the importance of vigilance and the signals his body might send. But on this day, such warnings felt distant, unanchored from the pragmatic cadence of life he'd followed diligently.

Feelings of vulnerability emerged, unexpected in their intensity as he weighed each possibility. Javier, ever the strategist, tried to confront the escalation of these unknowns with the logical rigors of risk assessment so familiar from his work—but the worry permeated each attempt to think clearly.

With deliberate calm, he reached for his phone, contemplating whom to alert about this incursion into his well-being. The urge to dismiss the sensation as fleeting was seductive yet instinctively troubling. As he sat, striving to distill uncertainty into understanding, myriad emotions cascaded through him—bewilderment mingled with a budding apprehension.

His heart, stirred with a desire to move past this anomaly, pulsed resolutely against the challenge. Javier's gaze fell upon the computer screen, the source of his morning's endeavor, now a tableau of emotion interacting with enterprise—two forces vying for dominance within him.

For now, patience and caution remained his allies as he sought to process what lay beneath the surface. The tapestry of this day—woven with both achievement and unease—unfolded with promises unkept and questions unasked, leaving Javier contemplating the complexities nestled within the everyday pursuit of vision and wellness.

The walls of the office seemed to close in around Javier, every inanimate object silently observing his unfolding struggle. His legs, the foundation upon which worlds had been built, refused service as he willed them to respond, to lift him with the ease that had always been taken for granted. Yet now, their rebellion was total.

The realization that he was not merely tired but experiencing something far more profound encroached upon his consciousness, gnawing at the edges of reason with a severity he could scarcely fathom. Anxiety seeped into his veins, its clammy touch accentuating the panic that began to surge.

Determined but incredulous, Javier leaned heavily on his desk, summoning every ounce of will into the treacherous limbs beneath him. The effort felt Herculean, draining him with a rapidity that was terrifying and unfamiliar. His breath came in ragged bursts, each inhale a struggle, each exhale unsatisfying. The oxygen felt both abundant and elusive, a cruel juxtaposition that heightened his fear.

His heart, usually a silent companion, now beat with an insistence that bordered on violence, each percussion a reminder of his mounting struggle. The drumming filled his ears, matching the crescendo of fear threatening to overtake him. Javier attempted one trembling step forward, then another, his feet barely cooperating as he staggered toward the door.

With each tentative movement, the pounding of his heart intensified, his body teetering on the brink of collapse. Thoughts scattered in chaos—the rational juxtaposed with terror, hope with desperation. The fear of falling, of succumbing to the weakness that enveloped him, gnawed at the fragile remnants of his confidence.

Each step was a battle. The kitchen—his last refuge—loomed ahead, blurred and shifting. Just a few more steps... But then, the world tilted. His legs failed. The floor met him with ruthless force. A moment suspended in time; gravity embraced him with ruthless mercy as he crumpled to the floor, the impact jarring yet dampened by the numbness overtaking him.

"Help!" The cry erupted from between trembling lips, a plea punctuated by the shattering of composure he had fought to maintain. His voice, once firm and assured, now trembled in the air, carrying with it the full weight of desperation.

Alan, ever alert, recognized the urgency, springing into action with a haste fueled by both duty and fear. Heart pounding with its own fervor, Alan grabbed the phone, fingers flying over its familiar keys as he dialed for assistance. Calmness in chaos, clarity in crisis—his words carried a steadiness Javier sorely needed to hear.

"Help is on the way, boss," Alan assured, his voice strong, intentional—a lifeline amidst the chaotic din of worry. The promise of aid resonated, the implications comforting even as they confronted the stark reality of Javier's condition.

From his vantage point on the floor, Javier felt the room spin, uncertainty cocooning him in its grip. The edges of his consciousness frayed as tiredness, panic, and vulnerability swirled into a vortex only release might calm. What had begun as a promising day had transformed dramatically, the fragility of existence made starkly clear in the stillness between Alan's words and the wail of forthcoming sirens.

Surrounded by tension and awaiting salvation, Javier lay immersed in the confluence of noise and silence, hope and dread, each pulse bringing the possibility of balance back to an unsettled world.

The Ride of Uncertainty

The blaring siren pierced the air, its relentless wail a soundtrack to the urgency of the situation. The ambulance surged forward, carving a path through traffic with the insistence of flashing lights. Within its confines,

Javier lay on the gurney, the rhythmic compression of the wheels against the road mirroring the tumult within his chest.

A paramedic loomed over him, a figure of reassurance amidst the chaos, her presence a calm harbor in the storm of anxiety and fear. The interior was a symphony of sounds—the steady beep of monitoring equipment, the chattering of radios, and the persistent siren blaring above all—a cacophony that enveloped them, isolating the scene from the outside world.

"Has this ever happened before?" the paramedic asked, her voice steady yet laced with a touch of urgency. Her hands, deft and skilled, checked his vitals with precision.

Javier swallowed, his throat tight. "No... not like this." His own voice sounded foreign, weak, and unsteady.

Through the haze of fear, he felt exposed, the vulnerability digging into him with relentless persistence. This was uncharted territory—being thrust into the unknown in a manner more intense and frightening than he had ever faced.

"Do you have a condition we should know about?" she continued, her gaze fixed on him, holding him steady amidst the rolling waves of panic.

"I'm on hemodialysis," Javier managed to say, his voice a mix of hesitation and clarity—a stark reminder of the fragile line he walked daily.

"Okay," she nodded, conveying professional reassurance. "We're rushing to the hospital; don't worry, you're in good hands."

Yet, despite her soothing reassurances, Javier wrestled with the ghosts awakened by this ordeal—the specters of past battles fought, victories hard-won, now looming with renewed threat. Each rumble of the ambulance wheels reverberated with the rhythm of his insecurities, amplifying the fear of losing control in a world that had once seemed firmly in his grasp.

The familiar comfort of routine and the optimism carefully cultivated over recent days felt fragile under this encounter with uncertainty. Each bump in the road seemed like a metaphor for the turbulence within, as he was forced to confront the reality of being at the mercy of forces beyond his immediate command.

Memories of previous struggles flashed across his mind like echoes of forgotten storms, moments when perseverance had turned the tide, slowly eroding into the present frailty he now faced. The irony struck him; the person who had crafted solutions and paths for others now yearned for anchoring in the face of his condition.

But through the whispers of doubt, a new realization began to take root—this confrontation was not just the undoing of security but also an opportunity for courage multiplied by vulnerability. The fragility he harbored was the same that had driven innovation and connection, offering perspective and inspiration where it had once seemed a limitation.

As the ambulance sped through the city streets, carrying him toward medical aid and clarity, Javier held close not just the fear and uncertainty but the understanding that these moments of adversity were chapters yet unwritten in his unfolding story. And in the harrowing symphony of sirens and roads, the echo of resilience began to play its tempered notes, signaling that while this journey challenged him, it did not define him.

Javier lay in the sterile embrace of the hospital bed, the soft hiss of oxygen and the steady drip of the intravenous line a stark reminder of the gravity of his condition. The room around him was quiet, a stillness that seemed to amplify the rhythmic beeping of the monitor, underscoring the vulnerability that had enveloped him.

The door opened quietly, admitting Dr. Wood, who approached with a professional calm that conveyed both authority and empathy. Javier shifted his gaze to meet the doctor, attempting to brace himself for whatever news was forthcoming.

"Javier," Dr. Wood said, his voice calm but firm, "your dialysis isn't working as effectively as it should. Your vascular access is compromised. We need to act now."

The words hung in the air, each syllable resonating with the weight of their implications. Javier's mind raced, trying to fully absorb the scope of what was being said.

"We need to take proactive measures," Dr. Wood continued. "A kidney transplant has become necessary to ensure your well-being. We'll keep you here for a few days to stabilize your condition. During this time, you'll undergo dialysis here, and we'll list you on the kidney transplant list."

Javier listened, his mind a whirl of thoughts and emotions. The simplification of a complex reality into these few sentences made the ground beneath him feel unsteady.

"Once you're stable, you can return home and await a donor match," the doctor explained gently. "As soon as we find a suitable kidney, we'll call you, and you'll need to come to the hospital immediately."

As Dr. Wood left the room, a profound silence descended, broken only by the rhythmic beeping of the machines that kept vigil over his recovery.

The reality of the situation slowly dawned on Javier, bringing with it a torrent of emotions. Fear edged its way into his consciousness—fear of the unknown, of waiting, of the surgery itself—the same fear that had threaded through his experiences yet felt amplified against the backdrop of his frailty. Along with fear came frustration—frustration at the betrayal of his own body, at the stubborn independence he now had to cede to circumstances beyond his immediate control.

Yet amidst the churn of anxiety, a quiet voice whispered hope. It spoke of the possibility of life renewed, of freedom from the arduous ritual of dialysis, of a vision unburdened by constant caution. Still, hope felt fragile to him, teetering precariously on the edge of that which was tangible.

As Javier lay back, his gaze resting on the sterile, impersonal ceiling above, a sense of disorientation seeped into the core of his being. Gone was the burgeoning resolve; in its place swirled a vortex of doubt and vulnerability. The confidence that had intermittently sparked within him felt elusive, like whispers of a distant echo too faint to decipher.

Questions began to creep in, insidious and relentless. Did he truly possess the resilience required for this impending journey toward a transplant? Each unknown seemed sharper under the hospital's fluorescent lights, a glaring reminder of the precariousness of his situation.

In this sterile solitude, he felt an unnerving loneliness, compounded by the knowledge that his friends, though supportive, lived worlds apart, bound by the ties of their own bustling lives. The reality of his single status weighed heavily; an absence felt more acutely in this moment of personal crisis.

His thoughts darted to the pizzeria—a dream in incubation but still a vision not yet fully realized, its potential as fragile as the hope he tentatively held onto. Faced with these looming uncertainties, Javier felt adrift, untethered from the shore of assurance he had once known.

A dark space enveloped him, folding in with silence and introspection. The forward momentum he craved was halted, obstructed by the very fragments of doubt that littered his path. In this solitary moment, the road ahead felt long, shrouded in shadow, and Javier found himself questioning whether he had the strength to navigate it alone.

Dawn and Discovery

The morning broke through the blinds with an assertive brightness, the suddenness of it edging Javier forcibly awake as the nurse orchestrated a new day's routine. The blinds rustled open with a brisk sweep, ushering in the fresh light of early morning, filling the room with a hopeful brightness that juxtaposed against his inner turbulence.

"Okay, Mr. Javier," the nurse began, efficiently setting about her tasks with a practiced hand. "Time to freshen up and tidy the bed. Breakfast is at seven, and Cindy, your kinesiologist, will see you at nine."

Javier nodded, absorbing the rapid delivery of information, surprised by the ease with which the nurse navigated the chores and his questions. "Is Dr. Wood coming by to see me today?" he queried, his curiosity leaning heavily on maintaining a grasp of the day's schedule.

"Yes, at 8:30," she replied curtly, moving through her duties with a determination that seemed second nature.

As the nurse departed, the room settled into a semblance of calm, albeit briefly. At precisely 8:30, Dr. Wood entered, his presence bearing both the assurance of experience and the weight of the morning's revelations.

"Javier," he addressed with a measured tone, "you have to be grateful this was caught in time. You experienced a potassium intoxication that could have led to far more severe complications, even a stroke."

Javier felt the gravity of the words as they resonated within the confines of his vulnerability. "What did you eat?" Dr. Wood pressed, exploring the source of the dangerously elevated potassium levels.

Instantly, Javier began cataloging his recent meals, certain in his adherence yet blind to the mishap. "I didn't eat tomatoes, avocado, or bananas," he said, frowning in thought. "I ate some nuts..."

"What?" Dr. Wood interrupted, concern creasing his brow. "Never eat nuts, almonds, or peanuts again. They're high in potassium. Okay?"

Javier nodded, the realization crashing over him. "Okay. I'll be more careful."

Dr. Wood, satisfied with Javier's understanding, continued, "I'll handle the paperwork for your kidney transplant application. If all goes well today and you have a good night, you can leave tomorrow morning. I advised Cindy to take you for a walk and some exercises; she'll be here at nine."

The plan settled uneasily over him, and the transition from patient to day-to-day gradually took form. "Okay," Javier replied, surrendering to the prescribed path.

Left once more to solitude, Javier grappled with a mix of embarrassment and resolution. The simple oversight—those unassuming nuts—served as both lesson and catalyst, calling for heightened vigilance. In hindsight, his gratitude for the timely discovery was overshadowed by a sense of vulnerability—an acknowledgment of the fine line he treads.

Yet through the noise of his confusion, a whisper of newfound determination emerged—a resolve to navigate his delicate path with foresight and awareness, ready to seize the future one deliberate step at a time.

Steps Toward Connection

The clock neared 9 AM when Cindy entered Javier's hospital room, an enigmatic presence infused with warmth and professionalism. Her approach was characterized by a quiet confidence, highlighted by her striking beauty—hair a cascade of inky black, eyes an inviting brown, carrying a sparkle that seemed punctuated by something lively and magnetic. Of Hispanic descent, she bore the features of her heritage with proud grace, her every gesture fluid and purposeful.

"Good morning, Mr. Lopez. I'm Cindy," she introduced, her voice carrying a friendly note that broke the room's sterile silence. Javier's heart gave a small, unexpected leap as her gaze met his, exuding genuine interest—a look that could only be described as sparkling.

Her manner instantly set him at ease, yet simultaneously unnerved him with a nervous anticipation he hadn't felt for years. As she guided him from the bed for their walk, Javier found himself deeply aware of how easily their conversation flowed. The effect she had on him was both exhilarating and grounding, a dichotomy of emotions sparked by her presence.

Their shared journey began with tentative steps as Javier reacquainted himself with movement. Minutes stretched into a symphony of steps punctuated by conversation that transcended simple pleasantries. They explored the small confines of the corridor and then the stairs, each shared anecdote a stair-step toward familiarity.

As they walked, Cindy mentioned her own father's battle with illness, her voice soft but steady. "It's tough, but it taught me to appreciate the small victories," she said, her words lingering like a gentle balm on Javier's weary soul.

With gentle encouragement, Cindy led him along, intuitively balancing the effort against his recent tolls. Despite the physical difficulty of returning to motion after his hours confined to bed, Javier found both solace and subtle empowerment in their exchanges.

Back in his room, Javier settled onto his bed, with Cindy seated opposite, her presence like a beacon of ease and engagement. They spoke of his passions and the potential held within the pizzeria. He expounded his ideas—each narrative laced with excitement and hope.

Engagement and laughter sparkled between them like warm light between distant stars. "That sounds wonderful," she said genuinely. "I'd love to visit sometime."

"That would be great," Javier replied, feeling warmth infused in his words, laced with both sincerity and a hint of timid anticipation. Their connection, although budding, felt significant—a tether of shared humanity in a sea of uncertainty.

After she left, her departure infused the room with a complex solitude. How did one define what arose between them? Was it interesting?

Professional kindness? He pondered, the insight somewhat elusive. Yet what remained was the joyful effect that her presence imprinted on him.

Alone once more, Javier's mind began cycling through reflections, a marriage of thoughts and emotions he inhabited. The stay at the hospital, necessitated by his illness, bore the hallmarks of vulnerability—a call to action to embrace the reality of his kidney transplant wait.

He found himself nestled into the layers of his reflection, questioning and untangling. The business, while brimming with promise, remained unfinished—an unfinished tapestry needing attention and care to achieve fruition.

Yet, it was Cindy who ultimately occupied his thoughts—her inviting smile, the effortless way she drew conversation from him, and the untold stories lurking behind her sparkling eyes. Her image lingered like a pleasant and intriguing enigma.

In this moment of introspection, Javier realized his life was populated by more than shades of gray cast by difficulties. Instead, color erupted within moments of inspiration and connection, challenging him not toward closure but toward a deepened understanding of his journey's unfolding chapters.

Resolved and renewed slightly, he whispered softly to the quiet room, reaffirming to himself the promise of renewal: to engage with the uncertainty, to anticipate the moments commanding full attention, and to treasure the spark ignited within when least expected.

As night fell within the hospital room, Javier lay cocooned in introspection, replaying the day's events with the saturated vividness of a memory too fresh to fade. Despite the day's promise of connections, a shadowy despondence lingered—an uneasy guest in the fabric of his consciousness.

It was in those silent moments that his mother arrived, her entrance as much light and warmth as the day's was tinged with uncertainty. "Hi, Javier! How are you?" she asked, her voice soft and filled with maternal intuition.

Her presence radiated comfort and familiarity akin to the lullabies of childhood. A smile, tinged with a mother's infinite understanding and concern, graced her face as she crossed the room. She enveloped him in a hug woven with tenderness—an embrace only a mother's arms could shore up with such unspoken reassurances.

The two of them sat in the evening's embrace, their conversation a gentle stream navigating between everything and nothing. Javier recounted the doctor's revelation—the necessity of a kidney transplant—a disclosure that had sharply redefined his current reality.

As the words unfurled, he saw an imperceptible shift in his mother's expression, a softness yielding to the sharp edges of ache as comprehension descended. A single tear escaped, sliding down her cheek, silent but heavy with meaning. Love, fear, and a mother's helplessness tangled in that one fragile drop. As his mother's tears fell, Javier felt a pang of guilt. He wanted to reassure her, to tell her he was strong enough for this, but the words caught in his throat, weighed down by his own doubts.

"Let's go, Javi. We're going through this together," she assured him, her voice a wellspring of hope and determination. "You're going to be fine."

Despite her unwavering support, Javier felt the weight of his condition pressing anew. The burden of witnessing her suffering compounded his own, adding layers to an emotional ledger he felt himself sinking beneath. The thought that she had tirelessly fought alongside him, each silent battle etched into the tapestry of their shared journey, swelled into an oceanic turmoil within him.

In that intimate moment, the urge to let emotion spill over became increasingly difficult to contain. Yet, he hesitated, holding firm against the sob threatening to fracture the tender silence for fear of its impact on her already tender heart.

"Do you want me to stay here with you?" she asked softly, an offer steeped in love and a whisper of self-sacrifice.

Softly, Javier responded, "No, Mom. Thank you. Please rest well at home. I'll be going home tomorrow; we'll see each other at the pizzeria."

An understanding passed between them, palpable and reassuring—a mutual acceptance of the quiet strength girding them both as she departed. In her absence, Javier was left cradled by night's embrace yet entangled in thoughts refusing solace.

Alone, his heart ached from the knowledge that his struggles mirrored in her eyes, that the world he crafted around his ambitions felt insufficient against the debt of love she bore without question. A sorrowful realization—a simultaneous acknowledgment of resilience and brokenness—wrestled his spirit into a darkened corner.

As sleep eluded him, the night unfurling long and indistinct, Javier allowed himself to momentarily unwind amidst the shadows of his thoughts—a reluctant acceptance of the journey ahead, cradled by a mother's unwavering faith and the stark reality of a warrior facing an uncertain battle.

The hospital room lay shrouded in muted quiet, only the gentle hum of machines punctuating the silence. As Javier lay in his hospital bed, the hour inching past 1 AM, he found the elusive embrace of sleep continually slipping away—the weight of the day's events too heavy to allow for rest.

In the darkness, clarity beckoned through the simple act of writing, a way to give voice to the storm within. He reached for a paper and pencil, his fingers brushing over their mundanity—an anchor in the churning tides of thought.

Here, in the silence of these hospital walls, the night's shadows seemed to mirror his own—a reflection of the struggle swirling inside with relentless momentum. For while the world outside this window sleeps, he finds himself at the intersection of vulnerability and realization.

Today, he learned that despite every measure he took to safeguard against the unknown, this journey demands an acceptance of vulnerability he hadn't anticipated. What happens tomorrow—a day blessed with promise or shadowed by uncertainty—is a story only time will tell.

Sitting with Mom tonight brought into sharp focus the depth of love she anchors me with, a love matched only by the weight of the heartache reflected in her eyes as I shared the news of a transplant. Knowing my battles echo in her soul fills me with profound gratitude, yet carries an unsettling regret that I cannot soften her burdens more tangibly.

These past days have been a testament to life's unpredictable rhythms—a medley of connections as seen in Cindy's kindness, betrayal in my own body's failure, and aspirations held in limbo for a transplant yet to be unearthed.

As the night deepens, the questions remain manifold: Do I possess the strength to continue undeterred by doubt? Can I render change palpable through challenges that now seem insurmountable at times?

The doubts pull at my resolve, threatening to erase the light peeking through my personal and entrepreneurial pursuits. And so, the realization dawns that space is needed; comfort sought not in solitude but through guidance.

Tomorrow, I shall reach out to Laura. For though tonight's darkness weighs heavily upon me, I trust in her clarity to illuminate paths that elude me now. Herein lies my commitment: to seek the support necessary to rediscover and embrace the promise within.

Tonight, fear and hope sit side by side, refusing to let me rest. Mom's love anchors me, but doubt still lingers in the spaces between. Tomorrow, I'll call Laura. I need her clarity to help me find my way back to myself.

As Javier set the pen down, a new resolve unfurled quietly within. While clarity remained elusive and the shadows of doubt were far from vanquished, he remembered the strength sometimes found in leaning upon others.

CHAPTER 9

A Morning of Renewed Resolve

Pale dawn light crept into Javier's hospital room, wrapping him in the hush of another morning. The familiar ritual of waking to find the nurse bustling in, orchestrating the start of his day, had yet to commence. His eyes flickered toward the clock—a glowing 5:55 AM stared back at him, as if marking time's indifferent passage against the restlessness keeping sleep at bay.

As minutes melted into each other, marking more than an hour past the typical schedule, the absence grew more conspicuous, mirroring the unexpected lulls of hospital life. Breakfast, when it arrived at 6:30 AM, finally disrupted the suspension of motion; the wheels of routine ground softly to life again.

With a tray set before him, Javier ate with a mechanical rhythm, compelled more by necessity than hunger. Each bite, though simple in flavor, bore an unspoken promise of sustenance, cradling him in the mundane embrace that kept uncertainty at arm's length. His mind lingered somewhere else, caught in fractured fragments of sleepless thoughts.

By 8:30 AM, Dr. Wood entered—a figure etched with purpose, carrying answers and care in equal measure. "Javier," he began, unfolding the day's revised expectations with the finality of someone wearing seasoned confidence. "You have to stay one more day. We'll run some last tests to stabilize your blood pressure and potassium."

The words hit like a dull punch. He exhaled sharply, disappointment settling in like a weight he couldn't shake. His emotional landscape shifted—

a tableau stained with the grayscale of delay. The delay gnawed at him, a reminder of how little control he had over his body's needs. But what choice did he have? He could stew in frustration, or he could redirect that energy—turn the waiting into something worthwhile.

"How do you feel about the kinesiology treatment?" Dr. Wood inquired, expertly pivoting the conversation.

The flicker of renewed purpose surged as Cindy crossed his mind—a spark painfully absent from the room's washed-out palette. His eyes widened in revitalization, embers refueled as he quickly responded, "Great. It's very good."

The opportunity to see Cindy again held the weight of silver linings, the prospect quietly buoying his spirits.

Dr. Wood affirmed the decision with a reassuring nod, initiating the next steps. "I'll have Cindy take you for some exercises today."

"Okay, Doc," Javier replied, the words laced with an eagerness just beneath their surface. As the doctor exited, Javier felt an undercurrent of joy, a flicker-laden path contrasting against disappointment's shadow. The enthusiasm, mingling with Laura's voice in his memory, coaxed reminders of resilience—the echoes of her words shepherding him back to reorient: If he couldn't finalize seeing Laura today, he would redirect.

Those sentiments reminded him of their hard-learned lesson: focus on what lies within reach to influence.

Thus resolved, Javier nestled into the phone's familiar comfort, dialing Alan at the pizzeria—a beacon of the everyday channeled within pragmatic urgency. Alan's tone, uncomplicated and grounding, flowed through the receiver. "Hey, Alan. How's the shop?" "Same as ever, Javi. No fires to put out today." "Good. Listen, I need my notebook. Can you have someone bring it over to the hospital ASAP?"

Hanging up, he felt the kineticism of agency coursing through his veins—that elusive sense of autonomy reinhabiting its rightful place within his spirit. In the landscape carved by persistence stretched alongside uncertainty, it readied him to pour into the funnel, continuing the project that awaited his hands.

In these decisions and recollections, beneath layers of complex emotion, Javier found tethered lines of hope. The day's initiatives tasted like potential: tangible, savory—a feast fated to progress once more.

The door swung open, and Cindy strolled in, her energy cutting through the hospital's sterile stillness like a burst of fresh air. "Hi, Javi! Where are we going today?" she asked, her voice a harmonious blend of enthusiasm and familiarity that effortlessly drew a smile to his lips.

Feeling emboldened by her easy manner, Javier chuckled, soaking in the informal camaraderie. "To the hallway and the stairs?" he replied, injecting a playful suggestion into his tone.

"Okay," Cindy agreed, her laughter a gentle melody that echoed through the corridor.

As they ventured out, their steps became synchronized, a rhythm threaded through the engaging fabric of their conversation. Words passed between them easily, each exchange covering a tapestry of topics from light-hearted stories to musings about life. For Javier, the simplicity of their rapport was transformative—a balm against a backdrop still tinged with the uncertainty of hospital life.

Yet amidst the ease, a self-inflicted doubt whispered as he took stock of their familiar banter. Was this dynamic unique? Did Cindy, with her effervescent nature, extend such warmth universally, or was there a genuine connection that beckoned him to explore further?

Resolved to navigate this uncertain terrain of feelings, Javier decided it was time to act upon his curiosity. With the familiarity of their conversation firmly established, he maneuvered towards the heart of his thoughts.

"So... we're launching something special at the pizzeria—Amore Pizza. It's a fun cooking course for singles, couples... anyone who enjoys good food. You should check it out. Maybe even join us?"

Cindy responded with an appreciative nod, her expression remaining friendly yet reserved. "Okay, thanks," she replied graciously, leaving no clear indication of whether his invitation had left an imprint. Was she just being polite? Or was there something in her smile—a flicker of interest, a quiet invitation to hope? The only thing Javier felt was his heart pounding harder when she looked at him; he felt nervous, something he hadn't felt in a while. Her gaze was magical, like that of a snake charmer, and that attracted him to her.

As they continued down the corridor, Javier silently reviewed the exchange. Did he read into her response too much? In the echoing silence that followed a bold realization, he grappled with uncertainty—a challenge he actively weathered to extend the invite.

"Well," Javier thought to himself, bolstered by equal parts admiration and resolve, "it's what I had to do—take the first step. Beyond that, the rest isn't in my hands."

Their exercise progressed along well-trodden paths, Javier's resolve tempered by the understanding that genuine vision required both boldness and patience, a willingness to set out the first steps and let life weave the remainder.

Crafting Dreams into Reality

Javier sat propped up against the pillows in his hospital bed, his notebook balanced on his lap as he delved into the digital realm of MKTGAIPRO. The sterile walls of the room faded into the background, the vibrant interface of the platform taking center stage in his world. Anchored in determination, Javier was set on transforming today's inspiration into a campaign that would resonate far beyond the hospital's confines.

His fingers danced over the keyboard with a newfound ease, propelled by the clarity of his vision as he navigated to Magic Hooks—an app designed to craft headlines that would pierce through the noise and capture attention. His thoughts crystallized into focus, and he fed the idea machine his key concepts for both his eBook and the course, entrusting its algorithms with the artistry of language.

Moments later, the tool unveiled its suggestion:

"Master Creativity & Connection: Unleash Passion in the Kitchen with 'Amore Pizza' – Your Guide to Delicious Discovery Awaiting!"

The headline struck a chord within him, a symphony of words that captured the essence of what he hoped to convey—a celebration of culinary creativity complemented by the delight of shared experiences, all wrapped up in the promise of connection.

With this powerful opening amplified through words, it was time to bring visuals into play. Javier transitioned seamlessly to the AI Videos app, a versatile tool at his disposal that allowed imagination to take form through imagery. He conceived an arresting scene to populate his digital canvas, accentuating not just flavor and texture but the joy that arose amidst the cooking itself.

One shot showed laughter erupting as flour dusted eager hands, a playful battle unfolding across the countertops. Another captured the magic of dough stretching and rising—transforming under careful, joyful touch. A third captured the moment when pizzas emerged triumphantly from the blazing oven—an array of vivid toppings contrasting against the crisp crust, tantalizing to behold. Fresh ingredients lay artfully displayed, hinting at the culinary journey culminating in perfect harmony.

The last scene encapsulated the exchange of smiles and wine glasses raised in celebration, a visual ode to the convivial nature of shared meals and momentous achievement. Across these frames, the narrative of Amore Pizza unfurled, one rich with culinary artistry that wove together strands of tradition and modernity.

Satisfied with his creations, Javier combined these visuals with the inviting headline and all essential details—reservation links, payment gateways, and the specifics of the event. He selected a launch date two weeks out, allowing anticipation to marinate alongside the logistical preparations. The funnel, completed, awaited the moment it would capture imaginations and draw curious palates. As he finalized the funnel, Javier felt a quiet triumph. It wasn't just about selling pizza; it was about sharing a piece of himself—a story of resilience, creativity, and connection—with the world.

Concluding this phase, Javier adjusted the budget, funneling $300 into its Meta promotion—each dollar an investment in visibility and engagement, driving the momentum that arose from a single spark of inspiration to the immersive experience awaiting new audiences.

As dusk settled, he lifted the phone with a sense of accomplishment mingled with anticipation. Alan answered promptly, his voice a reassuring constant amid the flux of ideas.

"Hey, Alan, it's Javier. I've set everything up. The promotion is live, and the Amore Pizza event is set. Please make sure the staff is informed and enthusiastic about the news."

Alan's reply, though brief, was replete with encouragement, echoing their shared commitment to the journey they embarked upon together.

As the call ended, Javier reclined against the dimming room, a profound sense of satisfaction buoying him. What lay ahead was a tapestry of imagined realities, crafted from a solitary room yet destined to resonate deeply within the hearts of many. The days to come held the potential to transform moments of inspiration into communal discovery—a testament to the determination ingrained within his vision.

The soft morning rays imbued Javier's hospital room with a gentle warmth as Dr. Wood entered, his demeanor one of quiet satisfaction. The sense of closure was tangible, and Javier could feel the promise of release readily approaching.

“Okay, Javier,” Dr. Wood began with a voice textured in optimism. “Your exams have come back normal. You’re free to leave. Pack your things and remember to be diligent with your diet.”

A wave of relief cascaded over Javier, a smile blooming on his lips—a deep, genuine expression of gratitude. The prospect of returning to the familiarity and sanctuary of home buoyed his spirits, a comforting balm after the confines and trials of hospital life.

As Javier left the hospital, the world outside welcomed him with a fresh breeze scented with the early notes of spring. The hum of the taxi as he settled into the seat felt like a gentle lullaby, almost as comforting as the rhythmic beat of his own heart drumming with new resolve. The familiar streets rolled by in a symphony of colors, each turn unveiling landmarks that whispered promises of routine and hope, of a life waiting to be reassembled.

With each passing block, the towering hospital behind him receded into memory, replaced by the comforting embrace of neighborhoods he knew well. The sight of a local park, children playing as carefree spirits, stirred a longing within him—a desire to return to life more vibrant and whole. His senses, dulled by sterile walls and antiseptic scents, reveled in the richness of the outside world: the rustle of leaves in the gentle breeze, the distant rhythm of footsteps, the soft murmur of daily life.

Back at home, the hum of normalcy wrapped around him as he closed the door to his sanctuary. The familiar creak of the wooden floor beneath his feet and the quiet ticking of the kitchen clock—these were the sounds of home he hadn't realized he craved. Each sound and scent was a testament to the resilience he had built within the hospital walls—a resilience that now found its grounding in the ordinary, in the promise of days yet to come.

Seated by the window, Javier allowed himself a moment of introspection. Hospital stays, once endless corridors of uncertainty, had shaped him with lessons of patience and vigilance. The unknown had lost some of its shadows, replaced by the soft glow of understanding that strength often lay not in solitude but in communities formed and in leaning on others.

With the familiar sights of his neighborhood unfurling beneath the dusky sky, Javier embraced a newfound clarity. It marked not just a return to life but the beginning of a journey painted with resolve, punctuated by the hum of life he welcomed anew.

Once home, dissolved in the comforting silence of his space, Javier's gaze landed on his diary—a testament to his introspective journey nurtured during his hospital stay. Settling into the familiar contours of his armchair, he transferred the raw, insightful words from hospital stationery onto the pages of his beloved diary.

Turning to fresh pages, he penned reflections anew—thoughts borne from the crucible of recent experiences:

Back home, the hospital feels like a past life—one that taught me patience, vigilance, and the power of small victories. Each challenge is a reminder of the strength found in leaning on others and embracing the unknown. This marks a pivotal beginning for me.

Lessons emerged from each moment—a sobering reminder of remaining vigilant with my health, listening intently to my body's whispers, and navigating life with mindful choices. The bonds forged with individuals like Cindy rekindled an appreciation for human connection and the unexpected kindness that reconstructs broken fragments.

Through doubt, I have unearthed the quiet strength of patience. Retrospection reveals the power in controlling what lies within my scope, illuminating paths obscured by uncertainty. Reaching out to others, fostering connections, and embracing support have been the threads stitching together resilience.

These times have also accentuated an inherent resilience—a seemingly binding moment of isolation and the returning warmth from family and friends. In adversity's workshops, I have excavated reserves of inner fortitude and clarity—a foundation upon which to nurture the future.

In overlaying dark spaces with learned revelations, hope whispers that the human spirit's capacity to endure can authentically redefine one's life narrative. It reinforces my commitment to seek balance amid triumphs and trials, weaving meaning into the wider tapestry of existence.

Moving forward, I hold these lessons close, cherishing creation, connection, and clarity as vital pillars in the unfolding narrative.

As Javier closed his diary, a palpable silence enveloped the room, reflecting renewal and new beginnings—a chapter closed, yet simultaneously, the horizon welcomed an unwritten future laden with promise and potential.

CHAPTER 10
Seeking Guidance

Javier sank into the familiar comfort of Laura's consultation room, but the weight of the past few weeks clung to him like an unshakable shadow. The room, familiar and inviting, had been a steadfast confidant throughout his journey—a safe harbor to weather the turbulence of his thoughts and experiences.

Sitting across from Laura, her eyes full of empathy and openness, Javier began to speak, the monologue pouring forth with a blend of relief and residual apprehension.

"Laura, the past few weeks have been... earth-shattering, to say the least. Being in the hospital, the kidney transplant news—it was like the ground disappeared beneath my feet at a moment's notice. Those nights, filled with machines and sterile walls, felt like they would never end, the uncertainty looming large.

And in the middle of all this, there's the pizzeria, my anchor and my passion, yet it's tethered to so much uncertainty. I launched this ambitious event—the Amore Pizza Course. It's something I'm genuinely excited about and terrified of simultaneously. It's supposed to bring people together to celebrate, to taste, to laugh, but what if I can't pull it off? What if it becomes another string of missed opportunities woven into the tapestry of all the times I've fallen short?

Fear clings to me like an old habit. I've stumbled so many times that the idea of standing tall—of truly succeeding—feels almost unnatural. What if I fall again, unable to get back up this time?"

As Javier's words hung in the air, Laura absorbed the depth of his emotions, her attentive presence a balm to the distress he'd just unveiled. After a thoughtful pause, she responded, her voice a steady stream of warmth and wisdom.

"Javier, it's apparent that you've been through an intense transformation, a journey filled with not just challenges but also significant milestones of growth. You've faced fears head-on—acknowledging them alone is a victory. The hospital, the news of the transplant, has all tested your resilience, reshaping your view of stability while also revealing a core strength you might have overlooked.

Change often masks itself as uncertainty, yet within it lies immense potential. You have created something with the Amore Pizza Course that holds genuine joy and connection at its heart; these are values you've built consistently over time. "Failure isn't in the fall, Javier. It's in refusing to rise again. You've proven time and again that you are capable of rising, stronger and more agile each time. The act of creating this event and facing the campaign's challenges speaks not just to the potential of your effort but to the power of your belief—belief in what you stand for.

Perhaps what you're feeling is not the anticipation of failure, but the heralding of possibilities yet to be realized. Let them unfold, trusting in the foundation you've re-forged and the community you're cultivating. Trust yourself, Javier, in this embrace of tension and promise.

As Laura spoke, Javier felt a shift begin—like the turning of the tide. Her words wrapped around him with the dual warmth of understanding and fervent encouragement. The layered fears—fears of falling again, of uncertainty, of the unknown outcome of his venture—began to dissolve, leaving behind a simmering realization: resilience had become intertwined with his story, not just in the moments of rising but in the choice to rise.

In that room, beneath Laura's insightful gaze, an elemental change took place. Javier found himself nurtured by budding confidence in his own potential, ready to unearth more from what lay within and ahead.

As Laura spoke, her words flowed like a soothing balm, offering warmth and understanding that Javier desperately needed. Yet, beneath the gentle reassurances, his inner voice remained vigilant, a skeptic perched within the shadows of his mind. "Could he really trust himself?" it murmured softly, reminding him of the past failures that seemed unwilling to fade. Every attempt, every setback replayed like a reel, whispering doubts that tangled with the present.

Javier's mind waged a quiet war as Laura's insights wove their tendrils around his heart. Her affirmation of his resilience was undeniable, yet his inner critic clung tenaciously to the notion of inadequacy. It raised questions, projecting images of past missteps and fears of faltering once more.

"Yes, I've overcome before," he thought, grappling with the complexity of belief and self-doubt. "But what if this time it's different?"

The fear sat heavy, a shadow that seemed to loom large despite the optimism Laura inspired. Yet beneath the layers of doubt, a flicker of hope—a possibility that perhaps this time trust in oneself wasn't misplaced.

As he listened, Javier knew that rising above his fears would not be instantaneous. It would demand embracing the uncomfortable and challenging the narrative of his inner critic. But maybe, just maybe, the time had come to reimagine what he could achieve by accepting Laura's challenge—to trust in his ability to rise one more time.

In that moment of reflection, he resolved to find the courage to believe in himself anew, letting hope and fear dance together, ushering him toward a future not defined by the shadows of the past but by the light of potential.

As Javier strolled through the familiar streets leading home, the session with Laura reverberated in his thoughts—a harmony of optimism and introspection that resonated long after her words had settled into silence. Each step propelled him forward with the quiet momentum of clarity, the realization of his unfolding resilience, and the promise of future possibilities surging gently within him.

Wrapped in a blend of reflection and the ambient city sound, he was awakened from his reverie by the buzzing insistence of his cellphone. Fishing it from his pocket, he glanced at the screen, his heart skipping as the message from Alan came into sharp focus:

"First Day: 50 spots already reserved from 60 available. Great work, Boss," the message read, exuding triumph and warmth with every character.

Javier stopped in his tracks, sunlight casting his shadow long and wide upon the pavement, as if elongating the moment for emphasis. A profound disbelief surged through him, quickly replaced by a cascade of exhilaration. He retraced his steps, this time with an eager urgency, each movement a testament to the joy sparking within him.

Bursting through the familiar threshold of the pizzeria, the flurry of activity within enveloped him like a vibrant embrace. Every detail—the aroma of baking pies, the hum of conversation, the cheerful clang of kitchen tools—took on an animated edge, marked by the realization of achievement mingled with the potential for even greater growth.

Javier made a beeline for his office, his sanctuary where ideas took flight. Facing the screen displaying the confirmation of reservations, he absorbed the unfolding of success—a concept brought to fruition through effort, vision, and the aid of technology that cleverly bridged dreams and reality.

Eager to share the news with his mentor, Javier dialed Mike's number, his anticipation crackling through the line. "Mike, we did it!" he exclaimed, the enthusiasm barely contained in his voice. "Why is this happening? I mean, it's amazing!"

Mike's response was equal parts sagacious and encouraging, the warmth of his amusement evident. "That's AI doing its work, Javi—you've implemented a great idea. Combining your passions with AI is like being a cyborg!"

Though laced with humor, the analogy struck a chord deep within Javier—the synergy between his vision and AI, a transformative ally. But amidst success lingered the call for further aspiration, detailed by Mike's gentle prompting.

"Remember, use every customer interaction as an opportunity to introduce something more. Think about offering the personal class at $3,000. Just two sales could dramatically change your month."

Javier's inner resolve was fortified, buoyed by a clarity of purpose enhanced by a clear path forward. "Yes," he affirmed, the commitment palpable in his voice. "And I'm already prepping for the second edition of Amore Pizza. This time, it'll launch together with my book."

Mike's laughter was a vibrant acknowledgment. "You're a machine, Javi! Keep building."

As the call ended, Javier reveled in the symphony of his own thoughts, the strides from café to office to connections unfurling like a dance, the rhythm marked by intention and potential. Each action orchestrated propelled him further, and each moment acknowledged fed the narrative of his ascent—a melody stitched from aspiration, challenge, and triumph.

The path now illuminated, Javier envisioned the possibilities beyond, ready to harness the momentum of success and channel it into new avenues, nurturing the courage to not just imagine, but to continuously create. The journey, bold and nuanced, embraced both inspiration and inventive spirit—a testament to the dream repurposed into reality.

The Dawn of Amore Pizza

The anticipation in the air was palpable as the day of the Amore Pizza event unfolded within the pizzeria. As the morning light seeped into the space, the restaurant was transformed into a vision of elegance and culinary promise. Every detail was meticulously arranged—the soft glow of candles flickered against polished wine glasses, the scent of fresh basil and wood-fired dough mingling in the air. The tables, adorned with crisp linens and twinkling fairy lights, whispered of an evening steeped in romance and indulgence.

Javier surveyed the room, his gaze taking in the rows of tables dressed in fine linens and the aromatic abundance of fresh ingredients displayed in delicate arrangements on plates. The aesthetic was quintessential—a symphony of color and promise, amplified by the inviting aroma of flour mingling with spices, ready to spark imagination and creation.

Alan arrived with aprons freshly pressed and bearing the event's brand, "Amore Pizza Course by Javier Lopez." Together, he and Javier placed them neatly over the chairs, lending a personal touch that whispered joy and pride in equal measure. The finishing floral touch—a union of white and red roses positioned invitingly alongside bottles of high-quality wine—cemented the scene as a beacon of intimate celebration.

At the heart of it all, nestled in the back, was the wood-fired oven—a humble hearth alight with readiness, symbolic of the culinary alchemy poised to transform possibilities into pizzas.

With everything double-checked and reassured by the visible preparations, Javier turned to Alan. "Alan, the guest list—how many did we have to leave out?"

Alan's response held excitement chased by genuine regret. "Two hundred, Javi."

The number jolted Javier—half incredulous, half exhilarated. The thought that many had wished to partake in this shared experience underscored the magnitude of their endeavor.

"Alright," Javier continued, resolve underscoring his voice. "Enter MKTGAIPRO and create a new email invitation with a fresh date in two weeks, offering a discount for those who missed out today. Let's add the benefit of arriving an hour early for a personal presentation of my book."

Alan nodded, grasping the urgency and potential of the situation. "I'll get on it. And what about a speaker for the book?"

"I'll look too," Javier agreed, considering the weight of this request. Expanding the day's focus offered depth beyond plates and dough; it carved an ecosystem of shared stories layered with purpose and connection.

As Alan departed to weave technology into mobilized action, Javier was left to absorb the charged atmosphere. With the restaurant nestled in readiness, a profound sense of effort and expectation beckoned that evolution he had long anticipated and prepared for.

In the quiet before the guests' arrival, Javier marveled at the energy coalescing around him. The confluence of creation, collaboration, and genuine excitement was a testament to a renewed narrative—a milestone in his journey shaped by intimacy and belonging.

As the minutes turned in quiet reflection, the clear vision remained: the event beyond a gathering was an invocation, an orchestration of people and experiences spun from a shared desire for celebration and connection. Javier's heart quickened as he realized anew the beauty of what lay ahead—a day knitted with love's flavors waiting to burst into being.

As evening descended upon the pizzeria, the room gradually swelled with life and energy, guests filling the space with anticipation and excitement. Couples, groups of women laughing and chatting, and a handful of men eagerly seeking new experiences converged, each buoyed by the promise of something uniquely enjoyable on this special occasion.

The warm glow of candlelight danced against the polished tables, casting playful shadows that flickered across smiling faces. The room was alive with a symphony of joyful laughter intertwined with the enticing crackle of the wood-fired oven, its flames whispering secrets of warmth and comfort. The rich aroma of basil mingled with melted cheese, wafting through the air and wrapping the guests in a sensory cocoon of culinary delight.

The tables were adorned with elegant floral centerpieces, the subtle fragrance of fresh blooms merging with the intoxicating scent of freshly baked pizza. As guests whispered eagerly over glistening glasses of wine, the soft clinking of cutlery and the gentle hum of conversation created a melodic backdrop to the evening's festivities.

Javier stood amidst it all, absorbing the vibrant energy that pulsated around him—every sensory detail a testament to the immersive experience he had crafted. This was more than a celebration; it was a tapestry of flavors and shared moments, a night etched in sensations that lingered long after the final dish was served.

Stepping forward, Javier addressed the assembled guests with genuine warmth. "Welcome, everyone, to Amore Pizza! This pizzeria has always been a family affair, one lovingly started by my mother and nurtured over the years. Tonight, we gather not just to make pizza but to make memories."

A nod from Alan signaled the servers to begin pouring wine, vibrant notes filling glasses with an invitation to relax and revel. Raising his own glass, Javier led the first toast, a gesture steeped in the bonds of camaraderie. "To Amore Pizza, where the love put into each pizza is only matched by the joy we share in every slice."

Applause swelled, and laughter lingered softly within the company—a resonance of togetherness marking the start of the evening's festivities.

"I leave you now with our talented chefs," Javier announced, seamlessly handing over the spotlight to the dynamic duo of the kitchen. The chef and sous chef, animated by a playful synergy, took center stage, joking their way through the creation of a classic Pizza Margherita with all the flair of seasoned performers.

Their witty banter and step-by-step demonstration drew in the guests, who mirrored their movements, hands covered in flour and delight. Amidst the aroma of fresh tomatoes and basil, the air crackled with the easy dance of radiant enjoyment—a distinct invitation for casual fun.

Without warning, the chef launched a cloud of flour straight at Javier, grinning like a mischievous child. Gasps turned to laughter as Javier retaliated, sending a white puff into the air. Within seconds, the kitchen transformed into a battleground of airborne flour and uncontrollable laughter. A cheer erupted as Javier, feigning shock, retaliated toward the sous chef, escalating into a flour-laden exchange.

Laughter echoed off the walls, the ensuing flour war drawing everyone closer. Guests joined in, flinging powdery dust into the air, mingling joyfully with the heady camaraderie shared by all. Amid the revelry, cameras flashed, capturing moments of effervescence for posterity.

Winding down, the chaos transitioned seamlessly into culinary creation as guests crafted their masterpieces to be whisked away to the awaiting oven. The crescendo of creative energy dovetailed into the indulgence of sharing—the ultimate culinary fulfillment realized.

Later, under quieter tones, the sous chef, Tom, tapped Javier gently on the shoulder. "Hey, Javi, see that woman back there?" he inclined subtly. "That's someone special. She hasn't stopped glancing your way all night."

Javier's curiosity piqued as he scanned the room, feeling his pulse leap as he confirmed Tom's observation. Cindy—graceful, enigmatic—was seated across the room among two friends, her presence a muse awash in candlelight.

A desire for composure gave him pause. He retreated to his office, allowing himself a moment of calm before he turned to greet her.

He approached Cindy, offering a greeting of familiarity—a nod to the nature of their shared moments. “Hi! So great you could make it,” he said, punctuating it with an affectionate kiss on each cheek.

“Congratulations,” she replied with a radiance that seemed to warm the surrounding air. “It's been very fun.” Cindy smiled, tucking a loose strand of hair behind her ear. "So, this pizza-making date…" Cindy’s lips curved into a playful smirk. "Do I get my hands dirty, or am I just here to admire the expert at work?"

Javier chuckled, a warmth spreading through him at her playful tone. It awakened something unfamiliar yet exhilarating—hope. "Only if you promise not to outshine me," he replied, his heart light with possibility.

His heart swelled at her words; a rich validation cradled in her sincerity. “Thanks,” he replied genuinely. “Can we talk at the end?”

“Of course,” she agreed softly, her understanding underscoring their exchange with promise.

Returning to his role as host, Javier immersed himself in the closure of the event, buoyed by the synergy of accomplishment and the potential for exploration. As glasses were raised for a final toast amid contented smiles, the evening’s success left an indelible mark, both on the fabric of his enterprise and on the evolving landscape of his personal growth.

As the event began to wind down, Javier felt an effervescent joy coursing through the room—a testament to the shared success and collective energy that had marked the evening. The steady hum of conversations, punctuated by laughter, echoed like a melody closing on its final, celebratory note.

Stepping forward, Javier gathered the room’s attention one last time. “Thank you all for being absolutely amazing tonight—not just participants, but remarkable cooks in your own right," he began, genuine warmth infusing every word.

With a subtle nod, he signaled his team, who unveiled a large, captivating gigantography measuring 1.5 by 1.5 meters. The striking image bore the words "Javi at Home Personal Classes," standing proudly at the room's focal point.

Javier continued with an inviting tone, "Before you depart, I'm thrilled to present a final offer exclusive to tonight. Imagine having me and my team at your home, your office, wherever you choose. We'll spend the day crafting pizzas that spark joy, culminating in a personalized creation, uniquely yours, added to our menu for an entire month!"

Faces across the room transformed with surprise and delight, an immediate ripple of excitement turning individual curiosities into contagious enthusiasm.

"The experience typically costs $4,000, but for those who book with Alan tonight, it's available for $2,500. Gather your closest twenty, and let's create an unforgettable culinary celebration together!"

Applause erupted, warm and vigorous, a percussion underscoring the vibrancy of the opportunity offered. Guests exchanged exuberant glances, some already lining up to see Alan, eager to grasp the chance of this exclusive event.

Javier watched, a deep and satisfying pride filling him as the scene unfolded—a tableau resonating with positive affirmations and infectious energy. Faces echoed anticipation and delight, infusing him with a sense of achievement interwoven with gratitude.

As guests approached to personally congratulate him, expressing admiration for the night's success, Javier basked in the ambiance of shared joy. Each handshake and nod buoyed him further, solidifying his belief in the journey now laid out before him.

A lively buzz filled the pizzeria as Javier moved through the room, the energy of recent successes charging the air. Amid it all, a woman approached him, buoyant with excitement. "I've never been able to make pizza like this at home," she expressed, her eyes bright with possibility. "I need this class!" she confirmed, extending her deposit of $2,500, her trust laid bare in her expectant gaze.

As Javier accepted the payment, a profound sense of accomplishment washed over him. This wasn't simply a business transaction—it was a connection, a testament to his skills and the trust people placed in his culinary expertise. The class was more than just an opportunity to teach; it validated his journey, reinforcing that his work had a meaningful impact on others.

Each interaction like this one became a building block in his growing confidence, a reminder that his path was not only about making pizzas but about fostering experiences, about guiding others in their culinary quests. The woman's enthusiasm mirrored back to him a sense of empowerment, igniting a flame of inspiration as he envisioned a future rich with both personal and communal fulfillment.

In these moments, Javier stood buoyed not simply by individual accomplishment but by collective experience—a tapestry of moments forged and memories embraced. As the night transitioned into personal farewells, Javier absorbed the radiant light of triumph, secure in the knowledge that Amore Pizza marked not just an event, but a milestone on a path illuminated by community, creativity, and heart.

As the evening ended, the soft glow of candlelight flickered gently across the pizzeria, a final adieu to a night that had unfolded with joy and discovery. Guests bid their goodbyes, smiles lingering as souvenirs of the evening's warmth and connection. Javier lingered in the ambiance, savoring the success and vibrant energy that had infused the air.

Through the dimming light and the now tranquil hum of the space, Javier's gaze settled on Cindy, who stood apart at the end of the room. The soft gold of wine swirled gently in her glass, catching the ambient glow, while serenity painted her features—a picture of soft contemplation.

A surge of emotion spurred him forward, steps paving the path toward her with deliberate intention. Her presence—a beacon grounding him amidst the ethereal whispers of lingering laughter and fading music.

"Hi," he greeted softly, the singular word suffused with warmth and anticipation.

Cindy turned, the arc of her gentle smile inviting him into conversation. "Did you like the event?" he asked, infusing the moment with genuine curiosity.

Her eyes twinkled, touching the space between them with approval. "It was awesome. Truly! We could have a dinner party like this with people from the hospital. The head nurse was with me and loved the idea."

Javier's heart swelled with unspoken gratitude, the interplay of professional admiration and burgeoning affection a complex melody resonating deep within. "Awesome," he echoed, their mirrored delight enhancing the harmony.

With quiet confidence, Javier raised his glass, the liquid inside catching the subtle flicker of overhead lights. "Let's make a toast," he proposed softly, "for having you here today—it brought me good luck."

Her smile, radiant and assured, rendered the room's ambiance secondary. In that smile, the essence of reality and potential juxtaposed, stirring emotions that thrummed with an exhilarating blend of hope and vulnerability. Javier could recall the sensation of his heart answering with warmth, subtle yet profound—a soft pounding that synchronized with each breath.

Caught within the moment's deep allure, Javier spoke with heartfelt sincerity. "Would you give me your number? I'd like to go out with you another day."

For a moment, time stretched, each second heavy with unspoken possibilities. Then, Cindy smiled—soft, knowing—and handed him her number. His pulse quickened, anticipation surging like a wave that refused to recede. Cindy's response was reassuring, her tone imbued with a gentle readiness. "Okay," she said, offering him the connection he earnestly sought as she shared her number—digits both ordinary and infinitely meaningful in their significance.

Javier received her number, his smile a harmonious fusion of joy and purpose. "It was good seeing you today," he murmured, each word articulated with the authenticity of growing affection.

The lightest kiss brushed her cheek—a touch of tenderness vigilant in its grace. The moment held—a fleeting eternity marked by burgeoning promise.

As he watched her leave, a gentle reverie enveloped him—the romantic nature of the evening unfurling in his mind like an elegant melody composed of shared narratives and intertwined hopes. Javier stood amidst the quiet restaurant, his heart embracing the illumination of a journey transformed by shared steps and aspirations yet to come.

Hope and admiration coursed through him—a tender yet resolute flow carrying his thoughts forward, crafting a shared chapter infused with possibilities for connection, accompanied by the heartbeat of dreams interwoven with the melodies of life's grand adventure.

As the last remnants of the evening faded, Javier joined his team in the final acts of cleaning, each gesture a small offering toward returning the pizzeria to its pristine state. The camaraderie shared in even these simple tasks imbued the space with a warmth that transcended the night itself—a testament to the dedication and unity that had propelled them through the milestone event.

Turning to address his team, Javier's voice resonated with gratitude and emotion. "Thank you, guys. You've responded like the great professionals and family you are. This is just the beginning—the first of many exciting events we'll create together. The next one will be in two weeks, and we'll introduce my book first."

As his words rippled through the room, he paused to absorb the moment, his heart swelling with pride and appreciation for the support that had made this success possible. The shared smiles and nods of his team bolstered his confidence in the journey ahead.

With the closing of another transformative chapter, Alan approached with promising news. "Javier, we've got three customers for the high-ticket $2,500 personal class."

"Thank you, Alan. Keep up the great work," Javier responded, his spirits buoyed further by the positive reception to their new offering, the potential for future success glimmering on the horizon.

Upon returning home, the quiet allure of solitude and reflection cocooned him. Javier slipped into bed, the pages of his diary beckoning as a canvas to delineate the lessons and revelations cultivated through the experiences of these past days.

Tonight, amid the hum of excitement and echoes of laughter, the experience of this chapter crystallized within him, whispering truths long cultivated and new insights freshly uncovered.

The Amore Pizza event was not just an exercise in orchestrating a gathering—it was a testament to the power of collaboration and shared purpose. Witnessing the team rally together, each member contributing their unique gifts, reinforced his belief that success is interwoven with community—an endeavor where everyone's potential blends to achieve something extraordinary.

True success isn't just about profit; it's about creating moments that linger—flavors that tell stories, laughter that becomes a memory, connections that outlast the evening, not just within this pizzeria, but also within himself. Visualizing dreams and encasing them is a venture bold enough to mark each daring step they take.

The connections fostered, both professionally and perhaps personally, kindled a warmth that transcends business and becomes something infinitely richer, rooted in the simple joys of human connection and understanding. Cindy's presence tonight stood as a beacon to the evolving path of possibilities, merging life's mosaic into fuller hues.

The dedication to growth has been both humbling and empowering, each accomplishment a stone building the bridge between today's actions and tomorrow's outcomes. There remains much to unfold, with the promise of his book looming as another layer—a chance to not just share recipes but share insights woven through the narrative of perseverance and reinvention.

In embracing these pursuits, he finds himself not only shaping the world around him but drawing inspiration and fortitude from those nearby—friends, family, mentors. Tonight reinforces what he knows: resilience isn't solely about enduring; it's about flourishing in the face of the unknown, remaining open to change and committed to transformation, regardless of how intimidating the horizon might appear.

As he carves these reflections into existence, he closes this chapter with gratitude in his heart, welcoming what lies ahead with a spirit eager to create, to connect, and to continue redefining the possible into reality.

CHAPTER 11
A Path to Renewal

Morning sunlight spilled through Javier's office window, casting golden streaks across his desk. With a steady breath, he flipped open his laptop, fingers itching with purpose. Today wasn't just another workday—it was momentum in motion. His fingers navigated confidently to the MKTGAIPRO platform, where the Ultra-Fast Funnels app awaited his next venture into creativity and strategy. The hum of anticipation underscored his every move, a soundtrack to the assured preparation for what lay ahead.

In mere minutes, the app presented him with three refined funnel options, each tailored to enhance and capture the essence of Amore Pizza's success. Javier's eyes danced over the possibilities before settling on a funnel that prominently featured a photo from Alan's video of the last event—a snapshot teeming with joy, flour in the air, and participants' animated expressions of culinary delight.

Building upon this vibrant foundation, he meticulously crafted the next layer: promoting the upcoming event with a hint of intrigue—his book presentation—a fresh narrative promising both continuity and novelty. Each keystroke imbued the plan with clarity and appeal, directly engaging the personas he'd so accurately defined, resonating with their desires for community, creativity, and culinary adventure.

Riding the wave of momentum, a spark of inspiration guided Javier to expand further, envisioning an idea that bridged business and community.

Swiftly, he navigated to the city's official webpage, bookmarking a spot to propose a pizza gathering alongside the mayor and his staff, initiatives aiming to create spaces of unity and shared experience—a homage to the spirit of Amore deeply sown into the pizzeria's philosophy.

He locked in the appointment—two days from now, at 11 AM, just in time to set the stage before the book launch. The timing was perfect, but a flicker of unease lingered. Could he navigate City Hall's red tape? This wasn't just about a meeting; it was about proving that Amore Pizza was more than a restaurant—it was a movement.

Not pausing long enough for complacency to settle, Javier orchestrated further exposure, securing a radio interview with a well-known station aimed at further elevating the pizzeria's profile, drawing even wider attention to the upcoming gathering and the promises it held. These moves resided within the framework of a new reality he'd crafted—where vision aligned with action, each an echo of the clarity born from his transformative chapter in the hospital. His confidence had transcended hesitation and doubt, swelling through the sediment of experience, filtering resilience, creativity, and engagement into his approach.

The continued growth in revenue, directly linked to these strategic shifts, spoke volumes to his ability to navigate challenges with renewed vigor—success materializing not just in numbers but in the empathetic connection formed with his audience.

Javier leaned back with contentment, ready to greet not just another day, but a future brimming with possibility. He had created not just a marketing strategy, but a living tapestry of authenticity and engagement—a narrative eager to expand its horizons, steadily leaping from pages yet to be penned.

With plans in place and a horizon filled with promise, Javier welcomed the tasks ahead with invigorated energy—a heart steeped in ambition and reconciliation with vulnerability, paving the way toward whatever awaited beyond dawn.

The midday sun cast a warm glow through the pizzeria's office, painting delicate patterns of light across the walls as Javier reviewed the latest figures on his laptop. The numbers told a story of growth—a testament to the meticulous strategies and community-building efforts that had come to fruition.

The door swung open, admitting Alan, who entered with a buoyancy befitting the news he carried, a subtle excitement illuminating his features. Alan burst into the office, barely containing his excitement. "Boss, the phone's been ringing nonstop! Sales are up 40% in just three weeks." He paused, grinning. "And get this—Pizza Margherita? Selling 50% more than any other pie!"

Javier paused, the weight of Alan's words sinking in like a revelation. "Wow," he exclaimed, a broad smile stretching across his face, the realization of their efforts coalescing into something tangible. "Revenue is scaling beautifully. We're on track to sell 100% more this month than the previous two combined."

The statement hung in the air, heavy with satisfaction and confirmation of their shared aspirations.

An embodiment of considered leadership, Javier turned his focus to his team—those who had stood shoulder to shoulder with him, whose dedication fueled this exponential rise. "Great work, Alan. I want you to speak with the team, understand how they're feeling, and let's establish a sales percentage bonus for everyone, and a quality bonus specifically for the kitchen team."

The directive was clear, a recognition not only of effort but of community—a gesture echoing his commitment to the well-being of all who contributed their talents and hearts to the pizzeria's continued success.

"Let's discuss this in our next meeting tomorrow afternoon," Javier continued, ensuring the steps to come were defined and the dialogue open.

"In the morning, I'll be meeting with the mayor. Here's hoping it turns out beneficial for us."

Alan's eyes met Javier's, a thoughtful observation resting just beneath the surface. He could perceive a shift—one more than just professional momentum. He saw a transformation in Javier, a blend of emerging confidence and decisive leadership skills seamlessly merging.

Alan absorbed these changes, an unspoken understanding forming—a mutual acknowledgment of the journey now in motion and the path forward, bright with possibilities.

As Alan left the office, tasked with engaging the team with this fresh surge of optimism, he felt invigorated, encouraged by his boss's decisive actions and his trust in their capabilities. The anticipation of tomorrow's meetings carried a different weight now, buoyed by expectations and underscored by gratitude and shared ambition.

Javier exhaled, staring out the office window. Growth wasn't just in numbers—it was in the faces of his team, the laughter echoing in the kitchen, the calls flooding in for reservations. This wasn't just business. This was a movement. It was an artistry of leadership—inviting compassion, enabling empowerment, and ensuring that each step taken was one forward in unison.

In those moments, Javier understood that while trajectory alone was meaningful, it was the touchpoints along the journey—the people, the visions, and the resolve—that truly brought life to the narrative of their collective success.

Leading Through Accountability and Vision

The room was filled with anticipation as the pizzeria staff gathered, each member feeling the undercurrent of something significant about to unfold. Javier, standing poised and confident at the helm, radiated a presence that was both calm and commanding. His demeanor set the tone for the meeting, a subtle but profound shift from mere weeks earlier.

"Thank you all for coming together. I wanted to talk about our journey over the past months and the transformation that we've experienced," Javier began, his voice steady, resonating with the blend of humility and resolve that had become a central part of his leadership style.

He took a moment, allowing the gravity of his words to set in before continuing, "We've faced challenging times—no doubt about it. Times when it felt like we were treading water, unable to break free from patterns that weren't serving us."

Looking around the room, Javier posed a reflective question. "Whose fault was that?" Silence stretched. A young kitchen helper finally spoke, voice hesitant. "The team's?"

Javier shook his head gently. "No," he reiterated, his gaze warm yet earnest. "Whose fault is it?"

The tension mounted, palpable in the charged atmosphere. The head chef, Carlos, stood hesitantly, taking ownership. "I am, Javier. But I've had..." he started, before Javier cut in, his interruption as gentle as it was firm.

"It's not you," Javier reassured, directing the focus back to the heart of the conversation. "So, whose fault is it?"

The silence stretched, tense with expectation, before Javier confessed, "Me. I was the problem. I let fear slow us down. I hesitated when we needed to evolve, holding onto the past instead of pushing forward."

His admission, grounded and open, settled over the room like a transformative wave, turning accountability into a shared journey rather than a solitary burden.

"It was through the difficulty of my kidney failure—the toughest mission I've faced—that I realized: waiting for change isn't an option. We need to innovate, to reimagine our day-to-day operations. Our suffering stemmed from my hesitation in leading us into something greater. I feared altering what I had from my parents."

His voice, full of conviction, rose with renewed energy. "But we are going to infuse this place with love, good food, entertainment, and incredible experiences. Our interactions will come alive through our AI platform, MKTGAIPRO. Despite personal challenges, I am here, undergoing dialysis three times a week and awaiting a transplant, ready at a moment's notice. But I have faith in you. Alan and Chef Carlos, you lead a team that has shown resilience through our recent event."

Applause erupted, a spontaneous affirmation swelling from staff who recognized Javier's transformative leadership—his words deeply resonating with their own experiences and aspirations.

"I'm pushing forward, setting up a meeting with the mayor to explore opportunities to serve our community. This is momentum we will seize, fueled by our collective strength."

As applause subsided, Javier added, his eyes twinkling with encouragement, "And before you return to your tasks, let me recommend a book that has inspired me greatly: 'Extreme Ownership' by Jocko Willink. It's about taking control, leading fearlessly, and driving change."

Over a casual team lunch, the air buzzed with excitement as Javier and his team gathered around the large wooden table at the pizzeria.

Carlos leaned back in his chair, a grin spreading across his face. "Never thought I'd see a boss who actually shares the success. Respect, Javi," he said, raising his glass in a toast.

The others nodded in agreement, a chorus of goodwill and appreciation reflected across their faces. Around the table, a renewed sense of camaraderie filled the room, mirroring a shared pride in their accomplishments.

"It feels good to know our hard work is recognized," added one of the servers, voicing a sentiment that resonated with everyone.

Javier watched, his heart swelling with pride. Their response was more than just gratitude; it was a testament to the collective spirit that had fueled their success, reaffirming the bonds that made them more than just a team—they were a family.

Buoyed by communal support and a shared vision, Javier's unwavering resolve was palpable—a testament to his growth and enduring optimism. This moment marked the nexus of personal evolution and leadership, instilling hope not only in him but also within each team member who carried it forward.

With his address concluded, an unspoken thread of unity wove through the pizzeria, drawing everyone tighter into a story of possibility and renewed purpose. Together, they would transcend the status quo, igniting a legacy set to shine ever brighter, one crafted with shared hands and unified hearts.

In the stillness of his office, Javier sat poised at his desk, where the weight of past uncertainties had been replaced by a renewed zeal for purpose. The steady hum of activity beyond the door had become a symphony of progress rather than a cacophony of unresolved challenges.

As Alan entered, Javier turned with intent, welcoming the presence of someone who had been a consistent ally throughout this transformative period. The room, marked by the bright spill of morning light, anchored the conversation about the new avenues ahead.

"Alan," Javier began with a confident demeanor, reflective of the shift not only in the business's trajectory but also in his own as a leader. "We've got several priority tasks to get through. First, I've scheduled a meeting with the mayor regarding the town's transparency web page. It's incredibly important for building our community ties."

He leaned back slightly, his gaze steady yet relaxed, a testament to his experienced grip over the multifaceted nuances of leadership he had embraced. "Next, I need us to prepare some free invitations for my book presentation coinciding with our next Amore Pizza event."

Alan nodded, fingers already poised to note the directives, his respect and admiration for Javier evident in the subtle attentiveness with which he engaged.

"And let's also offer some added engagement," Javier continued. "I'll need you to urgently produce five free invitations to the book launch and Amore Pizza event, set for the same day. Additionally, 50 vouchers with a 50% discount on any small pizza, redeemable on any Monday this month, would be fantastic."

Behind each request lay a strategic intent, one that underscored Javier's evolved understanding of market engagement, incentives, and community appeal. Alan could sense the thoughtful layers underpinning these decisions—a plan meticulously engineered yet human at its core.

"On those Mondays," Javier added, his voice threaded with warmth and creativity, "I'll host a seminar on inventing a new pizza. Alan, I'll need you to calendar everything, and please set reminders to keep us on track 24 hours before each event. Tomorrow is pivotal at Town Hall, so I need to prepare thoroughly tonight."

Alan acknowledged the role, inspired by the momentum Javier had harnessed and propelled. Through every task and challenge, Javier had manifested the inherent qualities of leadership—empathetic yet resolute, visionary yet attentively present in the moment.

"Leave the invitations and vouchers ready by 9:00 AM so that I can collect them at 10:00 AM. Set us up for success, Alan," Javier concluded, a genuine smile illuminating his face—the kind that whispered of articulated dreams translated into dynamic action.

"If this meeting works out," Javier expressed quietly, his gaze steady yet hopeful, "we could become a central hub for the city's events—a place where everyone feels welcome. It's not just about the pizza, Alan. It's much bigger than that. It's about creating a sense of belonging, a vibrant heartbeat for the community."

As the weight of his words settled in the air, Javier found himself grappling with doubt, a familiar shadow that whispered the hesitations of what-ifs and potential failures. Politics wasn't his game. Could he truly illuminate his vision effectively, showcasing the community epicenter he dreamed of creating?

Taking a deep breath, he reminded himself that perhaps this wasn't about playing their game by their rules—it was about demonstrating his passion, his vision, and letting them see the integrity behind why it mattered. The stakes were high for both his business and his personal growth. This interaction stood as a defining moment—an opportunity to bridge aspirations with tangible reality, and the path to achieving his dream lay just on the other side of this conversation.

With these thoughts aligning his focus, excitement and determination dulled the sharp edges of doubt. Javier was ready to step forward, aligning his goals not just with ambition but with the authentic purpose that fueled his journey.

Alan nodded, recognizing the gravity of Javier's aspirations. The pizzeria had always been a labor of love, but this opportunity held the promise of transforming it into a cornerstone of the city's social fabric.

"This isn't just business; it's personal. It's about growth, about carving our place on the map. It's an invitation for the city to join us on our journey," Javier added, his voice carrying the weight of both dreams and determination.

The stakes were undeniably high, a poignant mix of personal growth intertwined with business ambitions. This meeting was set to be a pivotal moment—a test of resilience and a stride toward broader horizons, embodying Javier's resolve to elevate the pizzeria from a local treasure to a cherished city landmark.

As Alan departed, a sea of gratitude welled within him—an appreciation for Javier's journey toward confidence and comprehensive stewardship that had redefined what each day promised.

Left in the tranquil aftermath of their meeting, Javier felt a surge of contentment tempered with anticipation. He had learned to glean value from uncertainty, taking careful steps that mirrored a harmonious balance between guided intuition and deliberate strategy.

These moments had crystallized not just the complexities of business advancement but the deeper narratives of growth that accompanied them, resonating profoundly through each connected choice and purposeful collaboration. For Javier, every decision was not just a step forward but a resounding echo of perseverance intertwined with purposeful passion. Each task undertaken was not just a duty but an invested promise to weave the varying threads of vision into a coherent and connected reality.

As Javier returned home, the weight of the impending meeting with the mayor began to settle in his thoughts. It was uncharted territory—a crucial conversation with the potential to help weave his fledgling endeavors into the community's fabric. Sitting quietly in his favorite armchair, Javier allowed his mind to wander over the possibilities and complexities of working with City Hall.

The political arena came to him wrapped in layers of both intrigue and caution—a domain he perceived as operating under an unwritten code of exchanges and favors, each representation carrying a certain weight. "What's in it for you?" and "How can we mutually benefit?" were likely questions to keep in mind.

Politics was a game of exchanges—of power, influence, and favors. Javier wasn't naïve, but he clung to one hope: that a damn good idea, backed by action, could still break through bureaucracy. It was a dance of diplomacy and genuine intent, a balance he was determined to achieve.

His eyes fell upon his diary—a steadfast companion through trials and triumphs alike. The familiar weight of the pen was comforting as he began to transfer thoughts to paper, hoping to crystallize the lessons the last experiences had offered into tangible insights.

Success isn't just numbers on a spreadsheet—it's in the trust I've earned, the risks I've taken, and the community I'm building. Tomorrow, I step into a meeting that could change everything. And no matter what happens, I know exactly why I'm walking through that door. Tomorrow, I face a pivotal meeting with the mayor. Whatever the outcome, I walk in knowing exactly why I'm here, ready to showcase the vision that drives me.

CHAPTER 12
An "Unexpected" Turn of Events

The morning carried a quiet sense of purpose as Javier moved through the pizzeria, gathering the freshly printed invitations for his book launch and Amore Pizza event. Each piece of paper felt like more than just an advertisement—it was a symbol of progress, a tangible piece of the vision he was steadily building. Each piece of printed material in his hands felt like a tangible manifestation of plans crafted with foresight and purpose—a symbol of the strides he'd made toward reimagining the future for his pizzeria.

Clarity, woven with confidence, accompanied Javier as he stepped into the familiar embrace of the town, light dappling through the canopy of trees lining his path. His strides were purposeful, each carrying him closer to the day's greater agenda: a meeting with the mayor set to begin in two hours. The preparatory gap afforded him time to reflect and refine his approach—a luxury he appreciated amidst the brisk pace of entrepreneurial ambition.

As he walked, a stream of thoughts cascaded through his mind—spanning the topic of navigating the political landscape and the threads connecting transparency and mutual benefit. Whispers of old adages played at the fringes of his consciousness, reflecting the adroit chess game of favors and alliances that often characterized such processes. In this play of acts, Javier recognized an opportunity to transform transactional interactions into genuine partnerships, anchored by shared intent and community-centered impact.

Reflecting deeply, he allowed his thoughts to percolate, evolving strategic tactics that balanced advocacy with introspection. By the time he arrived at the Town Hall for his engagement, those ideas had nested, embraced by a confidence woven through his recent achievements and seasoned by the approach he intended.

As Javier stood before the imposing threshold of City Hall, a myriad of emotions swirled within him. Excitement coursed through his veins, but doubt lingered at the edges, whispering the familiar refrain of inadequacy.

This wasn't just a meeting. It was a moment of reckoning. A chance to prove—to himself more than anyone—that he was no longer at the mercy of life's uncertainties; that he wasn't just surviving, but shaping his own future. Was he truly ready? Or was he simply putting on a brave face to play a role he hadn't fully embraced yet? These questions gnawed at him, tugging at the edges of his resolve. Yet, beneath the layers of uncertainty, a deep-seated determination flickered. He knew that this meeting with the mayor was an opportunity to showcase not only his business acumen but the vision that had fueled his journey—a vision of a pizzeria at the heart of the community, a nexus of connection and belonging.

Taking a deep breath, Javier reminded himself that perhaps this wasn't about outmaneuvering traditional politics but about showing them why his dreams mattered—an invitation to share in the narrative he had tirelessly crafted and which he was ready to bring to life.

With the stakes clearly defined in his mind, Javier stepped forward, each stride a testament to the fusion of ambition and authenticity that marked the path toward his dreams.

As Javier approached the imposing steps of Town Hall, he was met with an unexpectedly bustling scene—a crew from a television channel, complete with journalists and cameras, eagerly awaiting the mayor's arrival for a scheduled appearance at 10 AM. Pausing subtly to one side, Javier took in the animated chatter and the organized chaos that often accompanied media events of this nature.

A glimmer of opportunity sparked within him—the spontaneous recognition of a potential to intertwine his own story with the unfolding narrative. With Monday's serendipity at play, he approached the crew with confidence, a newfound assertiveness imbued by recent successes fueling his interactions.

"Hi there," Javier offered with an amiable smile, looking to foster a mutual benefit. "I own a pizzeria nearby—I'd like to give you guys a 50% discount on your pizzas today. I understand the commitment your profession demands, sometimes requiring hours of waiting to conduct interviews."

The journalist, sensing both authenticity and advantage, responded warmly, "Thank you! We work with Opera; we'll definitely take you up on that later. Hold on here for a moment. What's your name?"

"Javier," he replied, feeling the thrill of possibility ripple through the air. The mention of Opera caught his attention, as admiration for her work had long lingered in his consciousness—her story of overcoming hurdles akin to an unspoken guide in his personal journey.

Five swift minutes later, the journalist, now joined by her crew, prepared for a live segment, calling out to Javier, "We're about to go live in five minutes; would you mind if we include you?"

Javier, taken slightly aback yet unflinchingly eager, nodded his assent. "Absolutely, that would be wonderful. I've always admired Opera."

As the cue light on the camera blinked to life, signaling the start of the broadcast, Clara, the journalist, began, "Hi, Opera, we're here at City Hall awaiting Mayor Johnson. Before that, a kind local, Javier, has generously offered us discounts on pizza for lunch."

The surprise in Opera's voice echoed across the connection. "What? I envy you! I love pizza; he sounds most generous. Is he there? I'd love to say hello."

Following Clara's gestured direction, Javier stepped forward, heart pounding in his chest, yet his smile was steady—genuine and warm. He wasn't just being interviewed; he was stepping onto a bigger stage than he had ever imagined. "Hi, what's your name?" Clara asked warmly as they approached. "Hi, I'm Javier."

With the world seeming to narrow down to the directness of her gaze, he found himself slipping into an interaction he had dreamed of but never anticipated. Opera's voice, familiar yet direct, reached through the connection. "Hi, Javier. Just a quick thank you for your kind offer to my crew here. Beware—they eat a lot!"

A shared laughter enveloped the moment before Javier responded, his words imbued with sincerity, "Don't worry, we'll treat them with love and our best pizzas."

Seizing the opportunity to voice what had long simmered beneath the surface, he continued, "Opera..." he ventured, hearing the patient inquisitiveness in her prompt reply, "Yes, dear?"

"This is a pivotal day for me. Meeting you, even virtually, feels like fate. Your journey to fulfill your dreams has inspired me greatly. I am battling kidney disease and awaiting a transplant while striving to steer my pizzeria toward success. Your story fuels my will to keep pushing forward."

Opera's response was suffused with heartfelt warmth. "You're going to make me emotional, darling. You represent the unsung heroes who courageously embrace life in America. How can I assist?"

"Opera," Javier said, his voice steady despite the whirlwind of emotions, "this might be bold, but... I'd love for you to come to my Amore Pizza event on Thursday. It would mean the world to me."

"We'll make a point to," Opera assured softly, her words a thread weaving into the fabric of support. "I hope your aspirations blossom, Javier. Thank you for this delightful conversation."

As the crew continued their conversation with Opera, stepping back toward their broader agenda, Javier found himself rooted in overwhelming gratitude for serendipity and sheer possibility, embracing his day. The instant connectedness with Opera left an indelible impression, boosting his spirits with their shared camaraderie and granting him the feeling of an unseen ally now championing his cause.

As Javier stepped away from the impromptu moment with Opera, a strange mix of exhilaration and nerves settled over him. The interaction with her felt surreal, a burst of encouragement and notoriety that buoyed his spirits. Yet, as reality recalibrated around him, the impending meeting with the mayor loomed ahead—a stark contrast to the whirlwind media moment he had just experienced.

He took a deep breath, reminding himself of the stakes at hand. It's one thing to bask in the limelight of recognition, but another entirely to translate that attention into actionable business success. Tonight wasn't just about business; it was about building something lasting, a cornerstone for the community he envisioned. He refocused his thoughts, knowing that this meeting was crucial for both his dream and his determination to create something real and impactful, right in the heart of the city he loved.

His heart swelled with the realization of transformative potential at play, for the narrative of his life had intertwined spectacularly with moments crafted unexpectedly from courage and connection. The sensation was one of unfettered exhilaration, painting intricate patterns of what was unfolding—a tapestry where triumph and persistence serenely intertwined, ready to animate the dawning future with hope and unfathomable potential.

As the Town Hall loomed ahead, Javier felt both anticipation and readiness course through him, a blend of nerves and confidence that characterized the prospect of a significant opportunity. The nearby group of media personnel alerted him to the activity simmering in anticipation of the mayor's arrival. As he hovered at the periphery, Javier was acutely aware of the synergy between chance encounters and the preparation he had invested in his ventures thus far.

His watch ticked over to 10 AM, and the flurry of activity heralded a punctual appearance by Mayor Johnson. Amidst the introductions shared between Cindy, Opera, and the mayor, the conversation trailed toward community engagement, a central topic anchoring the mayor's ambitions for fostering connectivity throughout the rapidly growing city.

Struck by a sudden alignment of his own ethos with the mayor's vision, Javier experienced what could only be described as a 'eureka' moment—a realization rich with potential that prompted immediate action. Seizing this newfound clarity, he approached the reception area and courteously requested, "Could I wait in the designated room until my 11 AM meeting with the mayor?"

Smiling, the secretary affirmed his appointment, "Yes, Mr. Lopez, you're early, but you can make yourself comfortable in the waiting room just opposite his office." Following her directions, Javier navigated the staircase, the journey lined with photographs depicting the mayor's various inaugurations and civic milestones—classic imagery of progress intertwined with leadership.

With each step, a latent wave of anticipation stirred nerves that sought to test his resolve. For a fleeting moment, the simplicity of his recent achievements began mingling with an internal narrative of luck, threatening to chip away at his burgeoning self-assurance. Recognizing this unwelcome thought for what it was—a shred of old self-doubt—Javier consciously halted it, staving off the negative energy. He knew more than ever that the power of his intention was grounded in more than serendipity; it was purpose and deliberate.

By the time 10:45 AM arrived, Javier was determined to fortify his mindset further. With time yet on his side, he stepped into the men's room seeking a pocket of space conducive to centering his focus. Discreetly locking the door, he faced the mirror, its reflective surface serving as both a canvas for projection and a catalyst for resolve.

He remembered Mike's teachings on commanding presence through power poses—a technique echoing Amy Cuddy's research, with movements designed to physiologically and psychologically bolster confidence. Assuming the stance prescribed, Javier rolled back his shoulders, lifted his chin, and let out a slow breath. Power poses weren't just about posture—they were about owning the moment. And this was his. It was the physical embodiment of readiness—a tool not merely to mitigate anxiety but to activate the presence and assertiveness he intended to wield during the upcoming interaction.

For those five serene minutes, his form did not waver, each moment impressionistically reinforcing assurance. As the hand on the watch ticked toward 10:58, Javier emerged from his concentration, fortified by the tangible sense of composure and conviction awakened with expectancy—a prelude to stepping into the forthcoming dialogue with authenticity and assertive potential.

Javier took a deep breath, letting the remnants of his power pose steady him as he exited the bathroom. His confidence was as palpable as the anticipation in the air, an energy ready to propel him through the impending meeting with Mayor Johnson. Each step toward the mayor's office was firm, symbolizing his resolve and the journey he had undertaken to transform his dreams into reality.

As Javier returned to the waiting area, in two minutes, Mayor Johnson passed by and said:

"Javier?" the mayor acknowledged, his voice punctuated by leadership honed over years of governance. "Give me a moment, and we'll talk."

"Of course, thank you," Javier responded, the genuine assurance of his voice echoing the confidence now firmly seated within him.

As he prepared for the pivotal discussion about to unfold, Javier understood that he stood at the nexus of opportunity and ambition. This meeting was a testament to his journey of perseverance and innovation—a narrative not guided by mere chance but navigated through conscious action, dictating destiny's script with deliberate intention.

As Javier sat in the waiting room, the moments seemed to stretch with the weight of anticipation. The mayor's office was a showcase of power and history, fitted with grand antique furniture that whispered stories of the past. The luxurious leather chairs stood on a lush Persian rug, their timeworn elegance adding gravity to the room. A large wooden desk commanded attention, its polished surface reflecting the room's stateliness. Shelves lined with books on politics and civic matters hinted at the breadth of knowledge amassed within these walls.

A meeting table stood nearby, large enough to accommodate discussions shaping the city's future. Five people were seated at a side table, their presence a mechanical necessity of lobby laws, ensuring transparency and accountability in the mayor's negotiations. Each face reflected a mix of inquisitiveness and professionalism, their eyes occasionally flicking toward Javier as he waited his turn.

The atmosphere was thick with anticipation, a mix of nerves and excitement coursing through Javier's veins as he sat down. This was more than a meeting—it was an intersection of intent, where visions could be shared and synergies found. He was here to leverage community spirit through culinary engagement, a vision he had poured his very soul into.

Breaking the ambient silence, the mayor himself appeared, a figure of commanding yet approachable presence. "Javier?" he called with a nod, his voice carrying authority befitting his station. "Come in."

Javier stood, smoothing the front of his shirt, each step measured yet determined as he moved toward the mayor's office. Inside, the melding of history and function greeted him—a room configured for decisions of consequence, where the weight of responsibility could be tangibly felt.

The mayor gestured toward a chair opposite his desk, its high back a reminder of the conversations it had framed. "Please, take a seat. Water, coffee?"

"Water would be great, thanks," Javier replied, accepting the offer gratefully. As an assistant placed two glasses before them, the mayor initiated, "First, congratulations. Opera spoke very highly of you. You're on your way to fame, I hear."

Javier gave a modest laugh, the warmth of the exchange cutting through his initial apprehension. "Dear Mayor, sir, that was quite a happy coincidence," he replied.

Laughter was shared, a bridge across authority and ambition. "Call me Steve," the mayor offered, an informal invitation that hinted at a kinship of visionaries.

“Alright, Steve,” Javier said, leaning forward slightly. “I’m launching my first book—‘Healthy Delights for People Who Love Pizza and Have Chronic Kidney Disease.’ And after that, I’m hosting an event called Amore Pizza, where guests create their own pizzas in a fun, social setting. I’d love for you to speak at the book launch. I think your presence could make a real impact.”

Steve listened, nodding as Javier continued, "Also, I’d like to ask permission to use the central square in front of City Hall for a social pizza event. It's an opportunity for you to address the community on health initiatives, accompanied by special rice-flour pizzas for all to enjoy."

"We can make the city square more than just a backdrop for events. We can turn it into a community experience."

The mayor nodded, his interest clearly piqued. "And how do you plan to sustain engagement beyond a single event?" he inquired, his tone both curious and probing.

Javier didn't miss a beat. He had anticipated such questions, and his confidence shone through in his response. "With storytelling, consistent engagement, and collaboration," he replied. "We don’t just serve pizza—we create moments people talk about. Imagine a place where every visit feels like being part of a larger narrative, where the community's stories come to life."

The mayor paused in thoughtful contemplation, absorbing the potential woven into Javier's words. In this dialogue, both men recognized the opportunity to forge a unique partnership—one that transcended traditional business and tapped into the heart of community engagement.

This interaction, steeped in possibility, marked a definitive milestone for Javier, both in his personal growth and his business’s trajectory.

Javier's enthusiasm infused the room with a contagion of possibilities, his determination a palpable presence that commanded respect and attention. Here was a man bridging the culinary and civic in a symbiotic dance, where community engagement tasted as sweet as any culinary delight.

The mayor considered his approval a testament to shared values. "Sounds fantastic, Javier. As for Thursday, I'll aim to attend the book launch, though the Amore Party might be a stretch with my 6 PM prior commitment. The event idea is in the square I’m in. My marketing team will coordinate dates."

A surge of accomplishment filled Javier—validation that his dreams were aligning with reality. "Thank you, Steve. Here are tickets for the event and discounts for Mondays—though they’re for your assistant with the large family, I hear."

They parted with a handshake imbued with promise—an unspoken agreement of shared ventures and mutual benefit. As Javier left, buoyed by the encounter, his stride was infused with the energy of potential realized and further adventures awaiting in the tapestry of the city he cherished.

As Javier exited the city hall, his heart was buoyant with inexplicable joy. Each step toward his pizzeria felt light, yet his mind was active with puzzlement. Success had finally taken his hand after years of relentless effort. He couldn't help but ponder why things were beginning to flourish now when, in the past, despite his relentless dedication, results had been elusive.

He had poured countless hours into the pizzeria, obsessing over details, leading marketing campaigns, experimenting with new strategies, and yet, nothing had seemed to drive the breakthrough he was experiencing now. Was it the sophisticated AI at his disposal that made the difference? Or perhaps something else had shifted within him or around him. Was it about the timing, or had the world around him changed too?

With these questions swirling, Javier pulled out his phone. He needed to sort these thoughts and knew just whom he wanted to talk to. He dialed Laura, seeking a consultation after his dialysis session, hoping to untangle the complexities of this new chapter in his life.

As he continued his walk, the phone buzzed in his hand, interrupting his reflections. An unknown number flashed on the screen, and after a moment's hesitation, he answered.

"Yes?"

"Javier?" A warm, confident voice cut through his thoughts. "This is Astrid. I'm the head of nursing at the hospital. I was at your Amore Pizza event, and I loved it. Do you do private events for companies?"

Surprised yet intrigued, Javier hesitated for a split second. He hadn't ventured into that territory but seized the opportunity, echoing what had become his new mantra—embrace all opportunities. "Yes, absolutely," he replied with newfound assurance.

Astrid sounded pleased. "Could you send me a quotation for our consideration? We're thinking of an event for around fifty people."

"Of course," said Javier, his mind already briskly crafting the possibilities. "Please send me your email via WhatsApp, and I'll get that to you today."

"Thanks, Javier," said Astrid warmly, and with that, the call ended. Javier pocketed his phone, feeling a thrill of excitement at this unexpected turn. It seemed doors were opening for him everywhere, and he wondered again why now, of all times, everything seemed to be clicking into place.

He paused to reflect on the difference, finding solace in the realization that perhaps the change wasn't in the magnitude of his effort but rather in the essence of his approach. He hadn't simply applied himself—he had started showing up differently. In the past, every action had been calculated, anchored in a need for control and certainty. But now, he was allowing himself to adapt, to take risks, and perhaps most importantly, to open himself up to possibilities instead of clinging to control.

Every opportunity seized and every connection made had become woven into the fabric of this new phase of his life. Success wasn't just about persistence; it was about timing, about showing up differently, about embracing change instead of resisting it. Maybe the world hadn't changed—maybe he had. This understanding filled him with a deep sense of calm and readiness for what lay ahead.

He told himself it didn't matter so much why, but rather what he did with it now. As he approached his pizzeria, he shared the news eagerly with his team, his voice brimming with excitement and optimism.

The air was filled with possibility, and though he couldn't pinpoint the exact reason for this sudden success, Javier didn't need to understand it completely to appreciate and act on it. His session with Laura would help with that.

"I am going to dialysis now; call me if you need me. I'll go home afterwards and be here tomorrow morning. Today was for reveling in the fruits of our labor, the rewards of believing and persisting against all odds."

CHAPTER 13
Revisiting Hell

The sterile hum of the hospital ward wrapped around Javier as he lowered himself into the reclined chair, its worn cushions familiar yet unyielding. Machines droned in mechanical harmony—beeping, whirring, watching. Sentinels of survival, yet indifferent to suffering. Javier took a deep breath, smelling the faint antiseptic odor that seemed ever-present, a sterile aroma of cleanliness that felt like grasping at a semblance of safety amid vulnerability.

He felt the cold bite of the needle as it pierced his skin, sending a dull ache up his arm—a sensation he had grown far too familiar with. His skin, usually tanned and resilient, seemed paler in the fluorescent lights, the needle a stark contradiction against his flesh.

Each pulse of the machine was a reminder of lifeblood flowing—a rhythmic symphony that tethered him to the present moment. The faint squeak of the nurse's shoes on the linoleum floor bid a gentle reminder that life buzzed around him, even as his own seemed momentarily paused in this suspended routine. Yet, it was the chair, an old companion, that cradled

him with its unyielding cushions, bringing an odd comfort—constant in a world of medical flux.

As he sat, his gaze drifted to the window, capturing snippets of the world outside—a life filled with possibilities and promise, contrasting with this reality. He visualized the streets, bustling with the energy he longed to reclaim, a tableau teeming with the vibrancy of lives lived beyond these walls. Each landmark he imagined passed like a comforting whisper—promises of routine, of mundane beauty, of a life waiting eagerly to be reassembled beyond the sterile confines of this room.

The monotony was suddenly interrupted when his doctor approached, a shadow of concern evident in his features. "Hi Javier, how are you feeling?" the doctor inquired with an accommodating yet concerned tone.

Javier managed a weary smile. "A little tired, doc."

The doctor nodded, a hint of empathy in his eyes. "Yes, your dialysis sessions have not been as successful as we'd hoped. You will now need to undergo dialysis for four hours and thirty minutes each session. We're going to administer iron and erythropoietin three times a week due to your low red blood cell count. "The words settled over him like a lead blanket—heavy, suffocating. The antiseptic air suddenly felt sharper, stinging at the back of his throat. His stomach twisted, and for a fleeting moment, it was as if gravity itself had betrayed him, leaving him weightless in his own dread. Fear swelled within him, spreading like ink in water, darkening the fragments of hope he had painstakingly pieced together.

The doctor's voice grew softer. "You need a transplant soon, Javier. I hope it comes soon."

The words felt like a blow to his chest. His breath hitched, the room narrowing as his pulse pounded in his ears. He forced a nod, his fingers gripping the armrest tighter, as if holding onto something solid might anchor him in place.

With a slight nod, Javier murmured his thanks, attempting to muster the composure that now felt distant and elusive. The reality of his situation pressed down on him like a weight, smothering the fragile sense of stability he had been clinging to.

As the session continued, Javier sat quietly, surrendering to a storm of thoughts and emotions. The familiar fears of illness and the possibility of unmet dreams rose against him, specters from his past struggles, now as potent as ever. He was engulfed by an old, formidable darkness—a place where despair overshadowed clarity and vulnerability took the helm.

Despite the external quietude of the room, inside, Javier fought a silent battle against the overwhelming tide of what-ifs and uncertainties that clamored for his attention. Doubt nipped at his heels, urging him to question the foundation he stood upon.

As he watched the time pass, he grappled with the distressing truth of his situation. The medical setbacks seemed insurmountable, and yet a diminishing part of him whispered of resilience. But at this moment, darkness held sway, and Javier found himself struggling once more in the cyclic dance with fear he had hoped to leave behind.

Javier sat across from Laura, his usual poise giving way to an underlying tremor of anxiety barely held in check. The news from the doctor had shaken him to his core. As the silence in the room wrapped around them, it felt heavy and full, pregnant with unsaid words and feelings. When Javier finally spoke, his voice was tinged with vulnerability that was rare for him.

"Laura, I don't know where to begin. Every time I think I'm making strides, it's like life pulls me right back under. The dialysis... It's draining me. Literally and figuratively. Now they've increased the hours, and the treatments aren't as effective. I feel like I'm just stuck, just waiting. It's as if I'm in a tunnel that keeps getting longer, with no light in sight."

He paused, his hands trembling slightly, betraying the fear he battled continuously.

"I keep thinking about the pizzeria and everything we've worked for, and it just feels like it's slipping away. I feel alone again, like a child lost in the woods. My team is amazing, but it's as if I'm disconnected from my life and theirs. I'm a step behind in everything. And when I feel like this, I question if there's even a way forward for me. Maybe it's just fate," he said, a hint of resignation coloring his words.

Laura listened intently, her gaze steady and compassionate. She allowed the silence to stretch just a moment longer, understanding the weight it carried. Then, her words emerged, gentle yet firm, the kind that sought to bolster rather than rescue.

"Javier, anyone in your position would feel overwhelmed. That's not weakness—it's proof that you're still fighting. But listen to me—your fear, your exhaustion, they are real, but they do not define what happens next. It's okay to acknowledge them, to sit with them for a moment. But remember, where you go from here is still your choice."

She leaned forward slightly, her gaze encouraging. "You have shown incredible resilience. Think of how far you've come, even amidst obstacles. Your path may seem unclear now, but it's vital to focus on the steps right in front of you, however small they might seem. You're not alone in this journey. Let's look at what you can control, what decisions lie in your hands, and push one step at a time. Progress, no matter the pace, is still progress."

Laura paused, considering for a moment before speaking. "Javier, here's what I suggest," she said gently yet decisively. "Every morning, start by writing down one thing you have control over. Then, note one thing you can't control. Focus on taking charge of the first and let go of the second. It's a small practice, but over time, it can change how you see your challenges."

Javier nodded, absorbing her words. The simplicity of her advice was comforting, offering a tangible way to redirect his energy. Her suggestion was a lifeline in the swirling chaos of his thoughts—a strategy to help him regain a sense of power without feeling overwhelmed by what he couldn't change.

Javier said, "Laura," he began hesitantly, "what do you think has changed? Why am I finding success now when all my previous efforts seemed to lead nowhere? God knows I've worked hard all my life."

Laura leaned back, considering Javier thoughtfully. "Javier," she started gently, "I've seen how relentlessly you've pushed yourself over the years. But the key shift has come in your approach to uncertainty."

He listened intently, his curiosity piqued. "Earlier, your efforts were defined by actions driven by sheer willpower against the unpredictable tides of the market," she continued. "You've learned to manage uncertainty not by controlling every variable, which is impossible, but by adapting and responding effectively to the changes around you."

Javier nodded, piecing together her insights. "Think about how you responded to Astrid's call earlier. You didn't hesitate; you embraced the unexpected possibility and adjusted on the fly—that's confidence born of controlling what you can: your response, your mindset."

Laura smiled, her voice gradually shifting into that of a coach conveying a lesson. "Success isn't always just about hard work, Javier. It's about being receptive to change, seizing moments when they arise, and the mindfulness that allows you to see opportunities rather than obstacles."

In the glow of understanding, Javier realized the deeper truth in Laura's words—his triumph lay not merely in what he achieved today, but in how he navigated the unpredictable currents that had once seemed so overwhelming. With renewed resolve, he acknowledged that his journey was about continuous adaptation and growth.

As the evening deepened, Laura looked across the table, meeting Javier's eyes with a firm yet gentle gaze. "Javier," she began with conviction, "no longer doubt yourself—rise up. You've spent your life mastering the dance between the joys and sorrows that come your way. You've built resilience through everything life has thrown at you. Now, it's time to act on your intuition and experience rather than question them."

Laura leaned forward, fixing Javier with a steady gaze. "You've been questioning why things are finally falling into place," she began. "But I want you to dig deeper. What does success even look like to you now? Surely, it's more than just financial stability or staying afloat. If it's not about just surviving anymore, then what is it truly about for you?"

Javier hesitated, the question hanging in the air like a challenge. As he pondered, he started to unravel the complex emotions tied to his newfound achievements. It wasn't just about money or recognition; it was about creating a legacy, finding fulfillment beyond the day-to-day battles, and discovering a new sense of purpose that aligned with his dreams. He realized that success for him was now intertwined with community, innovation, and personal growth—a stark departure from what he had initially envisioned.

Laura's question unlocked a cascade of introspection, forcing him to redefine his understanding of success in a way that felt authentic and deeply personal.

She offered a reassuring smile. "You have an amazing opportunity on Thursday with the launch of your book and leading the Amore Pizza event. It's your moment to lead; trust your instincts, and I promise you, everything else will follow." Laura placed her hand over his, a symbolic gesture of belief. "Remember to also take care of your health. I truly hope the transplant happens soon."

Her words, clear and powerful, echoed in Javier's mind long after she had uttered them. As they said their goodbyes, Javier felt a surge of energy coursing through him—like a well had burst open, releasing untapped potential and vivacity.

The world outside seemed brighter, more inviting. As he stepped into a cab destined for home, he held onto Laura's encouragement tightly, feeling the vibrancy of her words rejuvenate his spirit. For the first time in a long time, uncertainty didn't seem like a daunting shadow; instead, it felt like a challenge he was ready to meet head-on.

As he lay down in bed, he took his diary and started writing:

"I always thought control meant security, that if I could predict every outcome, I could avoid disappointment. But control also suffocated the people around me—it made me rigid, unreachable. Maybe true leadership, in business and in life, is about trust. I've been so focused on orchestrating every detail and ensuring every piece of the puzzle stays in its place that I forgot about the importance of allowing others to take a piece and contribute their ideas and strengths. Letting go might be the key to building the kind of relationships I never knew I needed. It's time to embrace the unpredictability of life, to foster an environment where my team and my loved ones feel empowered to grow with me, not just by my side.

I used to believe that if I didn't control everything, it would all fall apart; that if I loosened my grip, things would slip through the cracks. But Alan stepped up. The team didn't just survive without my constant oversight—they thrived. And here I am, realizing that my need for control wasn't protecting the business—it was suffocating it. Embracing change and trusting others has opened doors I never imagined.

My mind often returned to the reflections on relationships I had with friends, family, and girlfriends. I thought about my penchant for controlled outcomes and how it shadows my every interaction. It's about control for me, isn't it? But here in nature, everything thrived not because it was controlled, but because it was left to evolve on its own. My job now is to embrace uncertainty as part of my personal growth, to foster relationships that breathe different air, freedom, and trust, and enjoy the path. Don't focus on the goal; enjoy the process.

Tonight, lying in my bed, I'm writing this to remember the lesson learned: Life isn't a series of tasks to complete but rather a journey that unfolds at its own pace. I need to let go of my timetable and trust in my ability to adapt when things go awry.

When he is writing, his phone rings. Javier sees the number, and it's Cindy. Javier thinks it's late; I'll wait till tomorrow, and he doesn't answer the phone. The phone stops ringing.

He sees the missed call.

Javier's phone vibrated on the nightstand, Cindy's name glowing on the screen. His thumb hovered over the answer button, but he hesitated, not out of indifference, but out of self-protection. If she truly wanted to be part of his life, a missed call wouldn't stop her. And if it did? Maybe that was answer enough. That could make her try harder and make her message clearer. He had too much to handle at the moment.

He told himself he'd return her call tomorrow, that by then he'd find the energy needed to engage in whatever conversation Cindy wanted to have. Yet, a small, insistent voice in the back of his mind whispered a cautionary tale—what if waiting until tomorrow wasn't enough? What if tomorrow brought a wave of realization or change that he wasn't anticipating? This thought hung in the air, hinting at the possibility of future interactions with Cindy that might carry more weight than he was willing to admit in this moment.

I'm closing this day knowing there are still pieces left to assemble. But now, I welcome the mess; it's part of life's puzzle. Tomorrow, I'll step forward—not with certainty, but with faith that I don't need all the answers right now. But tonight? Tonight, I let go. I let the stars remind me that some of the most beautiful things in life are the ones we don't control.

CHAPTER 14
Embracing the Unexpected

Javier arrived at the pizzeria before the sun fully claimed the sky, stepping into a world of controlled chaos—kneading dough, sizzling ovens, and the electric hum of progress. With MKTGAIPRO at his fingertips, he had transformed his marketing game—crisp videos, eye-catching graphics, and dynamic campaigns now dominated Meta, Facebook, Instagram, YouTube, and LinkedIn. The buzz was undeniable. Success had finally found its rhythm. As the dough for the day's pizzas began to rise, so did his excitement for the upcoming launch of his book.

Picking up his phone, he dialed his book editor. "Are we ready for the book launch?" he asked, his voice tinged with eager anticipation.

"We're all set, Javi," the editor replied. "I've sent you a case with 100 copies. By tonight, your book will be published on Amazon and Audible in both English and Spanish. It's a proud moment."

"Thanks, man," Javier said, feeling a swell of gratitude. "See you Thursday for the launch."

Just then, Allan entered Javier's office, a look of triumph on his face. "Everything is sold out for the event," he announced. "The sales in the last three months have multiplied fivefold. It's amazing, Javi."

"That's incredible, Allan," Javier replied, eyes gleaming with pride. "I need a hostess for the event and someone to help in the kitchen."

"Leave it to me," Allan said with a grin. "Let's make Thursday memorable."

As Javier noted down a few tasks, his thoughts drifted to Cindy. He picked up his phone again and dialed her number. She answered almost immediately. “Hi, how are you?” she said, her voice brightening his mood instantly.

“I’m good. I've just caught up with the work, but it’s paying off,” Javier began, his voice warm and engaging. “I couldn’t take your call yesterday; sorry about that.”

“Don’t worry about it,” she replied smoothly.

Javier, with a hint of flirtation in his tone, continued, “I’ve been thinking about you... and my book launch. Two great things that should be in the same place at the same time.”

Cindy could almost hear the smile in his voice, and she smiled back unknowingly. “Sounds like a fun evening. How could I say no to such an alluring invite?”

Javier chuckled, “Well, I guarantee you won’t regret it. Not only will there be an intoxicating mix of books and conversation, but the company? Absolutely exquisite. I promise.” “Alright, you’ve convinced me,” Cindy laughed, her voice teasing. “I’ll be there to see you shine.”

I recognize something in Cindy’s voice that steadies me—perhaps a tone of unwavering support I didn’t realize I needed.

As they ended the call, Javier felt buoyed by the conversation. Thursday was shaping up to be more than just a book launch. It was becoming a new stepping stone in his life, filled with exciting possibilities.

"If I could face my past without running, I could face anything—including the challenges ahead for my business."

It's Wednesday morning, and Javier is meticulously preparing for an important day at work. He is in his office, fine-tuning details for a big upcoming event, when Alan, his assistant, enters the room with an unexpected message.

Alan: Boss, there's somebody looking for you.

As Alan mentioned the visitor, Javier felt an instinctual tension before even seeing his father's face, as if bracing for a storm long predicted.

Javier: Who is it?

Alan: I don't know, an old man...

Javier: Ok, I'll be right there. Did you send the quotation to the hospital?

Alan: Yes, boss. They're good with it and want the event for two more Fridays.

Javier: Great. We need a crew trained for that. Please start the training right now—chefs, helpers, rent dishwashing machine, etc.

Alan: I'm on it, boss.

Javier steps out of his office, his mind still partially on logistics, as he approaches the reception area. There, he unexpectedly encounters an old man whose face stirs something deeply familiar within him.

The old man looks at Javier with a mixture of hope and uncertainty.

Old Man: Hi kid, so you don't know me now?

Javier's body turned to ice. His heartbeat slammed into his ribs, a staccato rhythm of disbelief and fury. The air in the room thickened, pressing in on him, trapping him in a past he had long since buried. He hadn't seen this face in years—the face that had once loomed large in his childhood, then disappeared like a ghost. His father. Emotions rushed at him like a tidal wave, anger leading the charge. Memories of abandonment and pain fought for his attention, but he was physically unable to move or respond immediately.

Javier's Thoughts: How dare he show up now? After all these years? What does he want from me? Why does he think he can just walk back into my life?

With effort, Javier summons the strength to respond, though his voice is edged with the emotion he's struggling to contain.

Javier: What do you want?

Father: I just wanted to see you... and talk.

Javier (still cautious, but there's a slight softening in his tone): We don't have anything to talk about. It's been too long.

His father looks at him, trying to gauge whether there's a chance to bridge the chasm that time and distance have carved between them.

Father: Please, just listen to me for a few minutes.

Javier stands there, the initial fiery anger slowly tempered by a stubborn unwillingness to show vulnerability. He nods, not out of agreement, but out of a sense of duty to himself to finally confront this ghost of his past.

Together, they delve into a complex conversation that will either pave the way to understanding or solidify the walls that divide them.

"I recount an incident during my military training, where I had to navigate a treacherous obstacle course in torrential rain. My boots drowned in mud, and my limbs ached with every agonizing pull-up. Yet, amidst the adversity, I discovered a resilience I never knew I had—a touchstone memory that shaped my fierce exterior."

Now I see you and I remember. In your early twenties, I stood outside your bedroom window in the house in Miami, hand poised to knock. The quiet sobbing from within paralyzed me, a stark reminder of my inability to provide comfort—a failure that lingered in every hollow attempt at connection.

I'm sorry, Javi; nobody taught me how...

"As his father spoke, images flashed in Javier's mind—a childhood where his father's stern silence loomed large against their sunny afternoons at the park, where his tentative laughter sought the man's approval. Today, he recognized the same stubborn resilience in his father that he saw reflected in himself, both a curse and a shared legacy."

Javier's internal struggle is palpable throughout this unexpected encounter. He is a man caught between his professional duties and the need to resolve the unresolved. Anger is certainly the first and most immediate reaction, but beneath it lies a river of emotions yet to be uncovered.

Javier: "I needed more than the math classes you gave me like in military school. I needed a hug or a word of confidence. You never showed what you felt for me."

Father: "I sent money every month. Your mother never told you?" His voice was sharp, defensive—like he expected that fact alone to absolve him.

Javier: "Yes, and do you think that's enough? You worked so hard in the military academy that you thought that's all I needed?"

"Do you know what my mother gave me at 14?" his father spat; memories visible like fresh wounds. "A ticket to a military academy. No hugs, no kisses. Just discipline and structure. I cried every night, and it was a sergeant who picked me up, not her. I became what I am because I had no choice."

His eyes, usually so cold, flickered with a distant sadness. "My father gambled everything away, and diabetes took him. I was left with nothing but a military scholarship and a mother too tired to care. That's it, Javi. That's why I don't have time for feelings. I don't really care. I had it worse."

Javier felt the weight of his father's words press upon him, a mix of empathy and frustration twisting inside. He had heard parts of this story before, but today it struck a deeper chord. Despite knowing every painful detail, the recitation still drove home the depth of his father's scars. The air between them grew tense, heavy with unmet expectations.

"Dad," Javier began softly, searching for understanding in eyes that refused to meet his own. "I know your story. Most of it. But I needed someone to teach me about life in other ways—not just through hardship and distance."

His father scoffed, arms crossed defiantly. The gap between them felt unbridgeable, two men on opposite banks of a turbulent river of pain.

Javier swallowed hard, his resolve firming. "It's too late now. I have my life to deal with." He pushed a slip of paper across the table, his heart pounding. "Here's my cell number. We can talk another day when there's more to say."

His father stared at the paper, anger and hurt warring on his face. But Javier had already risen, the moment too raw to linger over. He turned, feeling his father's gaze on his back as he wove through tables filled with laughter and chatter—a cruel contrast to the turmoil within him.

Once outside, the brisk air cooled his heated cheeks, a welcome relief from the tension inside. The door shut behind him, sealing the moment away. Javier took a deep breath, his heart aching but resolved. One day, maybe, they would find a way through this. But not today. Today, he needed space to breathe.

Maybe you weren't just my absence. Maybe you were the voice that told me to push harder, to rise when I fell. Maybe you never really left—maybe you just lived inside the part of me that refused to break.

Javier gathered the team with a sense of urgency in his voice: "Alan, get the team together. We have to talk before tomorrow." As they assembled in front of him, his demeanor shifted from urgency to reflective calmness.

Javier began, "I want to share a story that has inspired me deeply, one I hope will resonate with all of you. It's about Maya Angelou, a renowned writer and professor who stood before the world at President Bill Clinton's inauguration in 1993. She recited her iconic poem, 'On the Pulse of Morning.' Maya Angelou faced significant adversities, including racial discrimination and personal trauma, yet she rose above these challenges to become a celebrated author and civil rights activist. Her poem at the inauguration symbolized hope and resilience, resonating with many who have faced their own battles."

He paused, letting the weight of Angelou's achievement sink in. "This event wasn't just a platform for her literary voice; it was a symbol of personal courage—a defiant act of authenticity in a moment when many would falter. Angelou's recitation was more than words; it was a declaration of strength and individuality."

Javier looked at each team member with intensity. "In our project, as in Angelou's moment of triumph, we face the world with our voices, unfiltered and true. Tomorrow, as we unveil our work, remember her example. Let us speak with courage and authenticity, because that is our most powerful tool, our pulse in this collective journey."

He concluded, signaling with his presence as much as his words: "Be prepared, because like Maya, we are here not just to deliver our work, but to affirm our place, with all its responsibilities and glories."

"Tomorrow, we don't just serve food—we serve an experience. We create moments, laughter, and connection. This isn't just about cooking; it's about making people feel something. And I know—without a doubt—that every single one of you will shine."

As Javier concluded his speech, an electrifying charge filled the room. One by one, his team members began nodding in silent agreement. Alan, the ever-cheerful optimist, cracked a light-hearted joke, easing the tension and prompting soft laughter—a shared release of pressure and an unspoken agreement to move forward together with renewed determination.

This invocation of Angelou's courage and authenticity left a powerful imprint on the team, forging a collective resolve to not only complete the project but to transcend expectations, and Javier's leadership shone again.

CHAPTER 15
The Resilience of the Human Spirit

The restaurant hummed with anticipation, the low murmur of excited conversations mingling with the scent of fresh basil and warm dough. Guests streamed into the space, their faces alight with curiosity and celebration. The first dining area, set for fifty, glowed under the soft candlelight, its elegant décor a perfect prelude to an unforgettable evening. In its center, a large wooden table was adorned with an assortment of books that lent an air of intellectual curiosity, alongside two gleaming glasses of water meant for the hostess of the event, the Mayor, and Javier.

The room was packed, an indication of the event's significance and the esteem with which the community held the restorers. Beyond this initial gathering space, a second, larger dining room awaited more guests, ready to host an additional one hundred and thirty attendees.

Alan approached Javier, his expression a blend of excitement and anxious duty. "Boss," he said, "I've called a couple of local journalists; they're here to talk to you and take some photos." Javier nodded appreciatively. "Yes," he replied, "have any members of the opera crew contacted you?" Alan shook his head slightly. "They were here on Monday for lunch, but they haven't confirmed their presence tonight." Javier sighed subtly. "And the Mayor?" he inquired further. "His marketing crew is here, but he hasn't arrived yet," Alan confirmed.

As Javier engaged in conversations with local news outlets and magazines, weaving dialogues that combined charm and business acumen, he saw Cindy standing at the edge of the crowd. Her presence was like a beacon.

Javier's eyes found Cindy's across the crowded room, drawn to her like a current he couldn't resist. Moving toward her, he leaned in slightly, his voice warm with unspoken intent. "I was hoping we could steal a moment later... just the two of us." Cindy smiled back, her eyes sparkling with mutual understanding. "Yes, sure," she promised.

It was Javier's cue to nudge the evening into action. He approached the hostess, a poised woman whose confidence was matched only by her clarity of speech. With a gracious nod from Javier, she began the event with a voice that resonated through the room, capturing everyone's attention. "Good evening, distinguished guests, colleagues, and friends," she commenced, offering a welcoming smile. "We gather tonight in celebration of art, culture, and community. This evening, as we delve into shared experiences and future visions, feel free to engage, converse, and enjoy the essence of what brings us all together—our shared passion and commitment to supporting collaborative excellence."

Her words set the tone for the evening, a tapestry woven with enthusiasm, camaraderie, and a dash of suspense as the attendees eagerly awaited the mayor's arrival. The event, in all its grandeur, had officially begun, capturing the imagination and promising an evening of unforgettable experiences.

The hall buzzed with anticipation as guests took their seats, the ambient hum of conversation occasionally punctuated by laughter. A spotlight swept the stage, settling on the elegantly dressed hostess near the podium. With a welcoming smile, she stepped forward, adjusted the microphone, and began her speech.

"Thank you all for joining us tonight for the celebration of a book meant for the community and friends. We are honored to host the launch of this remarkable book that delves into improving the quality of life and honors people who have struggled with chronic disease and all the people who love them."

She paused for a moment to let her words resonate with the audience. "Our program will begin with insightful reflections from none other than our esteemed mayor, who has graciously taken the time to join us despite his demanding schedule. Following his remarks, our author, Javier, will share his thoughts before we move to the dining room for the 'Amore Pizza' event, where we will continue this evening's celebrations with good food and great company."

The spotlight shifted to the mayor, who had entered the hall with a blend of humility and authority, eliciting a round of applause. He gracefully made his way to the podium, extending a courteous nod to the hostess before taking the microphone.

A large banner read "Launch of '¿Kidney Disease? No Problem, 30 Exquisite Healthy Pizza Delicacies.'" A podium was set up in front of a gathering of community members, journalists, and healthcare professionals.

Mayor:

Walking up to the podium, adjusting the microphone.

"Ladies and gentlemen, distinguished guests, and champions of resilience, today we don't just celebrate a book—we celebrate a movement. A movement towards health, towards hope, and towards reclaiming the joy of food despite life's challenges. Javier's work is more than just a collection of recipes; it's a statement. A promise that no diagnosis can take away our love for good food, good company, and a life well-lived. Javier's remarkable collection, '¿Kidney Disease? No Problem: 30 Exquisite Healthy Pizza Delicacies,' is more than a cookbook. It offers a bridge between vibrant culinary tradition and the nuanced dietary needs of those facing the trials of chronic kidney disease."

Pauses, acknowledging the audience with a warm smile.

"As someone deeply involved with initiatives supporting kidney disease awareness and nutrition-anchored wellness, this work resonates with me. It fulfills a profound need for practical, enjoyable solutions in daily dietary habits. What Javier has crafted is a testament to resilience and creativity. It empowers not just patients but inspires all of us to view diet as a tool for joy and empowerment. Thank you, Javier, for enriching our community with this invaluable resource."

Javier

Rises from his seat amidst applause, approaches the podium.

The words from his manuscript flickered through Javier's mind like a cherished melody. Each chapter was a testament to his struggles, his resilience, and his hard-earned victories. He recalled the late nights spent hunched over his laptop, the periods of doubt and the persistent whispers of inadequacy. Yet here he was, standing on the cusp of a new possibility, the fruits of his labor within reach. The book was not merely a narrative; it was a chronicle of his evolution, a beacon that illuminated his path through the most daunting shadows. It was proof of how far he'd come and fueled his belief in how far he could still go. This, he realized, was not just his story but a manifesto of perseverance for anyone daring to dream beyond their limits.

"Mr. Mayor, thank you for your kind words and unwavering support. But tonight isn't about me—it's about all of us. It's about every person in this room who has faced limitations and still found ways to thrive. It's about taking back control over our health, our joy, and yes, even our pizza nights."

Turns to address the audience.

"This book is dedicated to everyone who has felt imprisoned by dietary limitations. It's for those who miss the simple pleasure of sharing a slice with family and friends. Our gatherings centered around food are more than meals—they are cultural stories, woven together from our past, meant to nourish our bodies and our connections. '¿Kidney Disease? No Problem' is my attempt to weave these stories with hope and health, one delicious slice at a time.

One of the first times I faced a strict dietary limitation, I found myself staring down my favorite food—pizza. Rather than giving up, I challenged myself to innovate. That's how I discovered the magic of cauliflower crust, something I thought I'd never enjoy. That moment taught me the value of flexibility, a lesson that resonates throughout my book.

As Javier recounted his story, a soft murmur of agreement rippled through the crowd. A woman in the front row nodded, understanding bright in her eyes. Nearby, a journalist scrawled notes furiously, while a young chef leaned forward, visibly inspired, soaking in every word.

He takes a moment, absorbing the emotion in the room.

"For those of us living with kidney disease, food doesn't always feel like a friend. This book seeks to change that relationship, to re-infuse meals with joy, community, and yes, deliciousness! Thank you all for being here to launch this journey of flavor and wellness. All the recipes are made for those who need a low potassium and phosphorus diet and want to eat great meals."

The audience cheers, and the event transitions to a lively book-signing and networking session, filled with conversations about recipes and health.

"I'm thrilled that so many of you are as excited about our new endeavor as we are," he said, nodding gratefully. "Your interest and support mean everything to us, and I can't wait for you all to dive into what we've created."

The hostess, who had been sitting at the front right of the table, stood up, signaling for quiet as he took to the stage. "Thank you, Javier, for sharing this incredible work with us," she began, a touch of pride in her voice. "Now that we've seen the wonderful new things brewing at Amore Pizza, let's continue this conversation with some good food and even better company."

With a smooth gesture, the hostess directed everyone's attention to the large oak doors at the side of the hall. They slowly opened, revealing the dining room filled with beautifully set tables, each adorned with candlelight and elegantly arranged floral centerpieces. The aroma of freshly baked pizza and artisanal dishes wafted into the room, inviting everyone to partake in the feast.

"Ladies and gentlemen, please join us in the dining room," the hostess continued, her voice welcoming and warm. "It's a night of celebration, and we are grateful to have you here with us."

Smiling and exchanging pleasant chatter, guests began to rise and move toward the inviting scene, eager to enjoy the culinary delights that awaited them. This evening promised more than just a presentation; it was a night to remember—a night that blended the creativity of Amore's new venture with the joy of coming together as a community.

As the afternoon sun cast a golden hue over the neighborhood, Javier stood proudly at the entrance of the event space, exuding an aura of quiet confidence. His posture was relaxed yet alert, reflecting the satisfaction of a job well executed. The mayor approached with a warm smile.

"Javier, this is extraordinary work," remarked the mayor.

"Thank you, Steven," Javier replied, his demeanor calm yet enthusiastic. "Are you joining us for the Amore Pizza event?"

The mayor shook his head regretfully. "I wish I could, but another engagement calls. However, my team will stay to capture the spirit of the event for our website and social media. They've also coordinated with your team for the central square city hall event."

"Great, thank you, Steven," Javier responded, appreciation evident in his voice.

As the mayor turned to leave, he threw a glance over his shoulder, a playful twinkle in his eye. "There's a surprise in store; you are having a prominent visit," he teased before hurrying away.

Caught off guard yet intrigued, Javier chuckled to himself, his eyes following the mayor's hasty departure. His mind buzzed with curiosity about the mysterious surprise. Pacing with a newfound sense of anticipation, he sauntered into the dining room. The atmosphere was lively; guests hovered around tables laden with culinary delights.

And there stood Cindy. Under the soft lights, she looked stunning—a vision that simultaneously calmed and aroused his senses. Javier felt his heart skip lightly. He greeted her with a confident smile, masking the flutter of excitement within. The event was in full swing, and Javier's heart swelled with pride. Everything was unfolding perfectly.

Across the room, Cindy watched Javier, her heart filled with her own blend of emotions. She had seen his journey, observed his fight from the sidelines, and marveled at his resilience. She felt a profound sense of pride swelling within her. Javier was more than just a restaurateur to her; he was a warrior, even if sometimes he didn't see it himself. Her gaze lingered on him, noting the subtle trembling of his hands—a sign of the turbulent sea of emotions within him. She realized her feelings for Javier were intertwined with admiration and a hope that spoke to their possible future—a bond tempered by trials but forged strong in resilience.

The evening was in full swing at Javier's restaurant, a vibrant blend of music, laughter, and the delicious scent of freshly prepared rice sourdough. The event, hidden from the casual passerby and reserved for a select group, was a wild success. Guests, eager to learn the secrets of their favorite dish, engaged with childlike enthusiasm, shaping and kneading dough while joking with one another, the air thick with camaraderie.

On the restaurant floor, Alan, the chef, and the sous chef caught the ripple of joy and moved their bodies to the rhythm, leading the rest of the team into a spontaneous dance. Laughter echoed against the walls as patrons joined in, a sea of swaying bodies and beaming faces turning the dining room into a festive parade.

Javier stood to the side, his heart swelling with pride as he watched the scene unfold like a beautifully choreographed ballet. Each member of his team played their role perfectly, their contributions blending seamlessly into a harmonious whole.

As he absorbed the scene, a sudden hush fell over the room, whispers rippling like waves as heads turned toward the entrance. And then, as if the moment had been scripted by fate itself, Opera walked in. Surrounded by cameras and an aura of effortless power, she took in the scene with a knowing smile, her presence commanding yet warmly familiar. The air crackled with energy. This was no longer just a book launch—it was a moment destined for history. The room collectively gasped in awe as Opera made her way through the crowd, greeting everyone with her trademark warmth and genuine interest.

Javier, surprised yet immensely honored, welcomed Opera with open arms, sharing a heartfelt embrace. Her presence elevated the night's significance, transforming it from a local celebration to something far-reaching and timeless. Opera gracefully ascended the makeshift stage, a natural-born narrator ready to share this special night's story with the world.

Taking the microphone, she scanned the sea of eager faces, her own expression a tapestry of admiration and sincerity. "Good evening," she began, her voice both commanding and comforting. "Tonight, we celebrate not just a business but a dream—Javier's dream—that has become a reality because of his unwavering passion and dedication."

I met this man some days ago, but it feels as if I've known him all my life. I felt identified with him, with his resilient spirit that I had to develop in my beginnings, when nobody believed in me, and I would have cherished some help. He is all about his community, and everybody likes him now!

She paused, allowing her words to settle over the crowd. "In these walls, we find more than just a restaurant," she continued. "We find a home, a place where every meal is made with love, and every customer is treated as family. Behind this success lies the heart of an anonymous hero—Javier—whose vision brings us all together in this joyful moment."

Her words, laden with emotion, resonated deeply with those gathered, each syllable weaving a connection, not only to Javier but to their own stories and aspirations. One by one, people lifted their glasses, a silent toast to the man who made a difference in their lives.

As guests murmured in awe, photographers captured the moment, and employees whispered excitedly about her presence.

The night continued with renewed energy, the attendees inspired by Opera's tribute, each heart brimming with admiration for the humble man who had created something truly extraordinary. Meanwhile, Javier marveled at the unexpected turn his evening had taken, overwhelmed with gratitude, aware that this was a moment he would cherish forever.

"Well," Javier grins, the tip of mischief dancing in his voice, "if I knew a pizza could make headlines, I'd have staged this event many years ago." A ripple of laughter courses through the crowd. "Oh yes, ladies and gentlemen, Opera wasn't just eating any pizza; she was imbibing art — the Lopez special, a culinary symphony crafted with as much passion as her own melodies."

Javier pauses, leveraging the air of spontaneity with the practiced ease of a maestro. "I must confess, getting Opera to visit us was more like a dream, and it came true, so thank you, Opera, for being here," he continues with a twinkle in his eye. "I was raised by my mother, who's here, so please, I ask for applause for her. Without her and my father's help, from a distance, we couldn't be here. I want to congratulate my team; they are awesome guys and professionals. But I want to say that today, we made a revolution in 24 hours. To do this, I've been working very hard all my life. Like the Disney cartoons I saw when I was little, you see the happy ending, but for me to have this day,

I've suffered and fallen and stood up again every time. So, raise your wine glasses, and let's make a toast for perseverance, change, love, and pizza."

A chorus of chuckles and appreciative nods affirms his words, the audience swaying under the warmth of his sincerity.

Javier steps back, the microphone settling safely into its stand as he takes a breath, soaking in the vibrant energy that pulses all around him. It's a moment that cements itself into the collective memory of everyone present, a scene of laughter, camaraderie, and dreams taking flight on the wings of what started as a humble slice of pizza.

The room was alive with the energy of the event, a vibrant hum resonating from the crowd as Javier Lopez continued to captivate his audience with his passionate speech. He stood confidently, the microphone an extension of his voice, commanding attention. As Javier stepped off the stage, the night air cooled his flushed skin, the echoes of applause still ringing in his ears. A sudden vibration against his palm cut through the laughter and clinking glasses. Javier glanced at his phone, expecting another congratulatory message—until he saw the caller ID. 'URGENT: HOSPITAL.' His breath hitched, the world tilting on its axis. The room around him faded, the celebration becoming distant white noise as he pressed 'Answer.'

Javier paused, a flicker of surprise crossing his features, microphone still in hand.

His phone buzzed with an urgency that eclipsed the routine calls of the day. His heart leapt into his throat, pounding with a relentless rhythm as he glanced at the screen. It was the hospital. A rush of anticipation mingled with a wave of apprehension. This was the call he'd been waiting for, yet a flicker of fear coursed through him. His years of waiting, of battling uncertainty, seemed to crumble into this singular moment of truth. How long had he imagined it? Yet, now that the moment was here, it felt surreal, as if he were standing at the precipice of the past and the future. Each breath was shallow, each beat of his heart a drum signaling hope clashing against dread.

The night had been a dream—success, recognition, a chance for something more. And now, fate had thrown him a lifeline. His heart pounded with a mix of excitement and fear. Could he dare believe this moment had finally come? As reality settled in, his excitement battled with the profound realization of what was at stake.

"Sorry," he said, slightly flustered. "This is unexpected."

Fumbling momentarily, he retrieved his phone from his pocket. The screen flashed urgently: "URGENT HOSPITAL." The room seemed to hold its breath; the very air thick with anticipation. Javier answered the call, his voice tentative but steady.

His voice came out barely above a whisper. "This... this is Javier."

"Yes. It's your kidney transplant. We have your kidney," the voice on the other end urged. "Take a cab, don't drive, and enter the hospital by the emergency entrance, please. ¡Come right now!"

The words hung in the air; their weight palpable. Javier's mind raced, absorbing the enormity of the moment. He abruptly ended the call, turning back to the startled audience, his expression a mix of excitement and urgency.

"Sorry, I have an urgent call for my kidney transplant," he announced, his voice tinged with disbelief and excitement. "My team will continue with you. I have to rush there."

With that, Javier stormed out, the crowd parting like the Red Sea. Their collective shock transformed into a symphony of supportive applause and murmurs of encouragement. Javier's footsteps echoed in the corridor as he sprinted toward the exit, heart pounding not only from the exertion but from the life-altering news.

The crisp air greeted him as he burst outside, frantically scanning for a cab, his heart a chaotic drumbeat matching the rhythm of his racing thoughts. Spotting a taxi, he waved it down with fervor.

"¡Hospital, please! As fast as you can," he urged the driver, settling into the backseat, his mind already racing ahead to the hospital, to the new future that awaited him.

The cab pulled away, leaving the event and his speech behind but carrying Javier toward a hope he'd long dreamed of—a chance for renewed life.

Javier burst through the doors of the emergency room, his heart pounding like a jackhammer. "I was called for a kidney transplant, urgent!" he shouted, his voice echoing off the sterile white walls. A nurse, efficient and calm amidst the chaos, approached swiftly.

“What’s your name?” she asked. “Javier Lopez,” he said. "Come with me," she instructed, guiding him through bustling corridors to a small preparation room. "Change and put on this robe to cover yourself," she said, handing him the thin garment.

"Please fill in these forms and take them with you."

Inside the room, Javier fumbled with the hospital gown, his fingers clumsy and trembling with adrenaline. The reality of what was about to happen was sinking in. Just as he tied the last string, the door swung open with a thud, and the doctor entered, all business and intense focus.

"Okay, Javier, are you ready?" the doctor asked briskly.

Javier nodded; his throat dry. "Yes," he managed to croak out.

"Are you allergic to anything?" the doctor continued.

"No, sir," Javier replied, his voice gaining a little more confidence.

"Alright, see you in a second in the pavilion. The nurses and assistants will take you there." With that, the doctor was gone as quickly as he had arrived, leaving Javier alone once more.

Soon after, a small team of nurses and assistants appeared, wheeling a gurney. As they transferred him onto it, Javier's heart thrummed like a wild caged bird, and the adrenaline surged anew. As they transported him through the sterile labyrinth of the hospital, he caught snippets of salsa music drifting through the air, a deliberate attempt to distract him from the severity of the situation ahead. One of them asked him, "Javier, I'll keep your phone, documents, and the forms you just filled in. Do you want me to call someone?" "Yes, please call my mother. Her phone is..."

In the operating room, the anesthetist leaned over him, a reassuring presence amid the sea of masked faces. "Okay, man, you'll start to count to ten when I tell you," the anesthetist said, his voice calm and steady. He paused for a moment, then asked, "What would you like to do when you're recovered from this?"

Javier smiled slightly; he blurredly saw Cindy, the tension lifting for a brief second. "I would like to make a revolution in 24 hours," he replied boldly.

As Javier lay in the hospital bed, he replayed the night's moments—the mayor's words, Cindy's smile, Opera's praise. It felt like a different life. One he was finally ready to step into. This wasn't just a second chance at life—it was a second chance to dream bigger.

"Alright, Javier, count from one to ten for me," the anesthetist instructed, his voice calm but distant. The bright surgical lights overhead blurred into halos as Javier exhaled, forcing his nerves into submission. "One..." His voice barely registered over the beeping machines. "Two..." A faint smile tugged at his lips. "Three..." He thought of his mother. "Four..." Of Cindy. "Five..." Of Opera's words. "Six..." The applause. "Seven..." The first bite of pizza that changed everything. "Eight..." His father's voice. "Nine..." The future waiting for him. "Ten..." Darkness.

Javier blinked slowly, trying to bring the blurred shapes into focus as he emerged from the haze of anesthesia. The gentle hand holding his felt familiar, comforting.

"Javi, your transplant was a success! Congratulations," a soft voice said. As he squinted, the face of Cindy came into sharp relief, her eyes warm with joy.

"Cindy? It's you?" he managed to whisper, a faint smile playing on his lips.

"Yes, it's me. But don't talk. I rushed here with Opera; she's covering the story outside. It's all gone viral!" Cindy replied, gently squeezing his hand. "I asked to be assigned to you. Now hush, save your energy."

Javier let his eyes close again, his heart swollen with gratitude. Hours rolled by in a mix of drowsiness and intermittent visits from nurses adjusting his morphine shots. Alone in his room, he finally received his phone. The screen lit up with an avalanche of well-wishing emails and, more surprisingly, five inquiries about business franchises for his pizzeria.

Javier made a mental note to start researching franchising models and to reach out to potential investors. He envisioned testing a pilot location to ensure that the heart and soul of Amore Pizza could be replicated without losing its essence. The journey of scaling his vision began here, at this exhilarating crossroads.

His mind buzzed with possibilities. "So many paths to consider," he thought, a swirl of emotions as potent as the pain medication coursed through his veins. The notion of rebuilding his relationship with his father surfaced, a challenge he felt ready to face.

"I need to know Cindy more," he resolved silently, sensing an unexpected but welcome connection. Thoughts of scaling his pizzeria into a franchise flickered through his mind, fueled by dreams of unprecedented expansion.

As another wave of pain threatened to breach his consciousness, another shot of morphine lulled him back, this time gently into a place of dreams where futures are built from murmurs of the past.

The future held many promises: a potential partnership with Cindy, an empire from a single dream, and perhaps, healing an age-old rift that whispered of family.

With a final deep sigh, Javier drifted into a deep sleep, comforted by the thought of new beginnings and alliances yet to come.

What will tomorrow bring? A relationship with Cindy? A thriving business or the rekindling of family ties? Only time would tell in the next chapters of his life.

To be continued....

The End.

About the Author

Sebastian Vasquez is a multifaceted professional with a background in Engineering and Business Administration, holding an MBA and a diploma in Marketing. As a certified AI Expert and Ontological Business Coach, Sebastian combines technological insight with business acumen to foster innovative solutions. Having thrived in roles as a Marketing Manager and Franchise Owner and navigating the challenges of living with chronic kidney disease, Sebastian embodies resilience and growth. He passionately develops new ventures and products with AI, driven by a life trained in overcoming uncertainty. Connect with Sebastian to explore paths of transformation and success.

Made in the USA
Monee, IL
04 November 2026

67387619R00095